SpringerBriefs in Petroleum Geoscience & Engineering

Series editors

Dorrik Stow, Heriot-Watt University, Edinburgh, UK
Mark Bentley, AGR TRACS Training Ltd, Aberdeen, UK
Jebraeel Gholinezhad, University of Portsmouth, Portsmouth, UK
Lateef Akanji, University of Aberdeen, Aberdeen, UK
Khalik Mohamad Sabil, Heriot-Watt University, Putrajaya, Malaysia
Susan Agar, ARAMCO, Houston, USA

The SpringerBriefs series in Petroleum Geoscience & Engineering promotes and expedites the dissemination of substantive new research results, state-of-the-art subject reviews and tutorial overviews in the field of petroleum exploration, petroleum engineering and production technology. The subject focus is on upstream exploration and production, subsurface geoscience and engineering. These concise summaries (50–125 pages) will include cutting-edge research, analytical methods, advanced modelling techniques and practical applications. Coverage will extend to all theoretical and applied aspects of the field, including traditional drilling, shale-gas fracking, deepwater sedimentology, seismic exploration, pore-flow modelling and petroleum economics. Topics include but are not limited to:

- Petroleum Geology & Geophysics
- Exploration: Conventional and Unconventional
- Seismic Interpretation
- Formation Evaluation (well logging)
- Drilling and Completion
- Hydraulic Fracturing
- Geomechanics
- Reservoir Simulation and Modelling
- Flow in Porous Media: from nano- to field-scale
- Reservoir Engineering
- Production Engineering
- Well Engineering; Design, Decommissioning and Abandonment
- Petroleum Systems; Instrumentation and Control
- Flow Assurance, Mineral Scale & Hydrates
- Reservoir and Well Intervention
- Reservoir Stimulation
- Oilfield Chemistry
- Risk and Uncertainty
- Petroleum Economics and Energy Policy

Contributions to the series can be made by submitting a proposal to the responsible Springer contact, Charlotte Cross at charlotte.cross@springer.com or the Academic Series Editor, Prof. Dorrik Stow at dorrik.stow@pet.hw.ac.uk.

More information about this series at http://www.springer.com/series/15391

Dayanand Saini

Engineering Aspects of Geologic CO$_2$ Storage

Synergy between Enhanced Oil Recovery and Storage

 Springer

Dayanand Saini
Department of Physics and Engineering
California State University
Bakersfield, CA
USA

ISSN 2509-3126   ISSN 2509-3134 (electronic)
SpringerBriefs in Petroleum Geoscience & Engineering
ISBN 978-3-319-56073-1   ISBN 978-3-319-56074-8   (eBook)
DOI 10.1007/978-3-319-56074-8

Library of Congress Control Number: 2017935958

Printed on acid-free paper

This Springer imprint is published by Springer Nature
The registered company is Springer International Publishing AG
The registered company address is: Gewerbestrasse 11, 6330 Cham, Switzerland

This work is dedicated to my Ph.D. adviser, Dr. Dandina N. Rao, Professor of Petroleum Engineering, Craft and Hawkins Department of Petroleum Engineering, Louisiana State University, Baton Rouge. It was Dr. Rao, who introduced me to the fascinating science and engineering of CO_2 injection for EOR purposes and laid down the foundation of my interest in topics like CO_2-oil miscibility.

Foreword

With the majority of the world's scientists with a background in climate change agreeing that human induced increases in the concentration of greenhouse gases in the atmosphere, with a particular focus on CO_2, are creating changes that exceed any known natural changes caused by regular cycles (sun output, other climatic cycles), there is a need to take action to reduce the rate of this increase in CO_2 concentration (see discussion in the IPCC Fifth Assessment Report 2013). As discussed in this report, one of the methods of preventing CO_2 emissions from reaching the atmosphere is to capture the CO_2, transport it to a storage site and put the CO_2 underground where it will not be impacting the atmosphere. The storage can be either direct injection into a saline formation (saline water with no economic value) or into reservoirs that contain hydrocarbons (enhanced oil recovery or EOR). In the latter case, assuming significant oil still in place, there will also be a production of incremental oil. The focus here is on the use of EOR to provide significant storage with an added economic benefit of oil recovery.

Major international groups, notably the International Energy Agency, have reported on the significant opportunities that exist globally for CO_2 storage through EOR. This might reach 360 Gt over the next five decades (IEA 2015). The economic benefit, besides reducing CO_2 increases in the atmosphere, would be the production of some 375 billion barrels of new oil. While it seems intuitively wrong to use the CO_2 to produce more hydrocarbons, a life cycle assessment suggests otherwise and that this is a valid method for CO_2 storage with environmental benefits (Suebsiri 2010). While this SpringerBrief book focuses on the United States, and it is certainly true that the majority of active EOR projects are within the US, it should be noted that other countries are increasing their understanding of EOR through pilot projects and a few commercial projects. Just as the first well was drilled in the US in 1859, the second well and the first to be put on production was in Canada (Williams No. 1 well at Oil Springs, Ontario), so the first EOR project to be researched for the safety and effectiveness of storage in an active reservoir was in Canada at Weyburn, Saskatchewan (Wilson and Monea 2004). This research was not an examination of the EOR process, but was designed to gain a better understanding of the storage of CO_2 in the oil field and to examine the ways in which the progress of the CO_2 could be monitored. The development of monitoring techniques would also provide a mechanism whereby any leaks into overlying strata could be determined if such leakage should occur. One of the findings of the research was that leakage into overlying strata was not occurring, suggesting that this method of storage was indeed safe. Currently, this IEAGHG Weyburn-Midale CO_2 Monitoring and Storage & EOR Project has stored more than 25 million tons of anthropogenic CO_2 and considered the world's largest carbon storage project currently in operation (Brown et al. 2017).

The purpose of this SpringerBrief book is to look at the engineering aspects of EOR and the associated storage that goes hand-in-hand with the incremental oil production. Dr. Dayanand Saini has done an excellent work for the concise summaries of cutting-edge research and practical applications of this important subject of our time. It examines the broad range of expertise and techniques that are required to effectively manage an EOR project, as well as to monitor what is happening in the subsurface and ensure the integrity of the storage. An EOR project has a clear mandate to make money for the investors in the project as well as to provide environmental benefits. It does this by minimizing the amount of CO_2 required to effectively produce the oil and to ensure that the CO_2, a costly solvent, is not lost to the atmosphere or surrounding strata at any stage of the process. Inevitably some CO_2 is brought to surface with the oil; this is captured, recompressed and reinjected in a closed-loop system. The need to be careful with the CO_2 also provides assurance of the safe storage of the gas.

While most of the CO_2 currently used for EOR globally is from natural sources, there is a growing volume of anthropogenic CO_2 being utilized. Indeed, all the CO_2 going to both Weyburn and Midale fields in Saskatchewan, Canada, are from anthropogenic sources (Dakota Gasification in North Dakota and the coal-fired Boundary Dam Electrical Generating station in Saskatchewan). To be effective, the CO_2 must be captured, unwelcome impurities removed, compressed and moved by pipeline to the EOR sites. This capture must be undertaken in a cost-effective manner to allow the economics of EOR to be realized and the benefits accrued to the environment (Idem et al. 2015).

We expect that this SpringerBrief book will be a valuable source of information for engineers, scientists, and decision makers working in the academia, industry, and government. We also hope that it will be a useful resource for the next generation of young researchers working in this important research area of carbon capture and storage.

Dr. Malcolm A. Wilson
Member of the International Advisory Board
Energy Academy of Europe

Dr. Paitoon (PT) Tontiwachwuthikul
Fellow of Canadian Academy of Engineering and Professor
University of Regina

References

Brown K, Whittaker S, Wilson M, Srisang W, Smithson H and Tontiwachwuthikul P (2017) The history and development of the IEA GHG Weyburn-Midale CO_2 Monitoring and Storage Project in Saskatchewan, Canada (the world largest CO_2 for EOR and CCS program). PETROLEUM (Elsevier) (in print)

Idem R, Supap T, Shi H, Gelowitz D, Ball M, Campbell C, and Tontiwachwuthikul P (2015) Practical experience in post-combustion CO_2 capture using reactive solvents in large pilot and demonstration plants (Review Paper). Int J of Greenhouse Gas Cont 40:6–25

IEA (2015) Storing CO_2 through enhanced oil recovery. International Energy Agency. OECD/IEA http://www.ipcc.ch/—the Intergovernmental Panel on Climate Change (IPCC)

Suebsiri J (2010) A Model of carbon capture and storage with demonstration of global warming potential, Dissertation, University of Regina, Canada. https://en.wikipedia.org/wiki/History_of_the_petroleum_industry_in_Canada

Wilson M, Monea M (2004) IEA GHG Weyburn CO_2 monitoring & storage project. Summary report 2000–2004

This book provides an introductory discussion of several key engineering aspects of geologic CO_2 storage, through the analysis of currently operational large-scale simultaneous CO_2-enhanced oil recovery (EOR) and storage projects, to make a case that the petroleum industry can embrace the CO_2 injection as a profitable service business. An enhanced synergy between CO_2-EOR and storage will not only help the petroleum industry in being competitive in low carbon economy, but it will also improve the public acceptance of geologic CO_2 storage projects. The main purpose of the book is to convey this message while explaining the key engineering aspects of CO_2-EOR such as CO_2 injection strategies, monitoring of injected CO_2, CO_2-oil miscibility, maintaining the integrity of cap and reservoir rocks and wellbores, selection of the favorable storage sites, and the use of modeling tools for making rational decision about the suitability of CO_2-EOR sites in a practical manner.

Chapter 1 describes the expertise and experience that petroleum industry has acquired over the years in operating both the commercial CO_2-EOR projects as well as employing the same for operating simultaneous CO_2-EOR and storage projects for achieving a balance in anthropogenic greenhouse gas (GHG) emissions by sources and their removal by sinks of GHG. Chapter 2 analyzes the key features of currently operational large-scale integrated carbon capture and storage projects (LSIPs) and other large-scale simultaneous CO_2-EOR and CO_2 storage projects in USA and Canada.

Chapter 3 describes the additional activities that an operator must include in a commercial CO_2-EOR project for transforming it into a simultaneous CO_2-EOR and CO_2 storage project. It includes the monitoring, verification, and accounting of CO_2 in both the injection and closure (post-injection) phases while establishing a baseline for existing subsurface, near surface, and atmospheric CO_2 concentrations. So far, petroleum industry has mainly used CO_2 injection for recovering the stranded oil left behind in depleted oil fields, however in the process a significant portion of injected CO_2 has also been stored in the reservoirs. An understanding of the reservoir engineering aspects of currently used injection strategies that are more focused on enhancing the oil recovery from the reservoirs can help the operators in devising new injection strategies that are more amenable to both EOR and CO_2 storage. Chapter 4 attempts to serve this purpose.

For achieving the synergy between EOR and CO_2 storage it is critical that CO_2 is injected in miscible mode. For designing a miscible flood, fast cost-effective and reliable estimate of minimum miscibility pressures (MMP) is necessary. Chapter 5 discusses the topic of CO_2-oil miscibility in a simple and practical language for developing a better understanding of this complex but one of the key designing parameter of a simultaneous CO_2-EOR and CO_2 storage project. Maintaining the integrity of CO_2 injection sites in commercial EOR projects is necessary for economic reasons, however maintaining the integrity of storage sites (injection zone (reservoir), overlying seals (caprocks) and the wellbore penetrating the caprocks and reservoirs) is the center piece of engineered containment of injected CO_2 in the subsurface. Chapter 6 provides the detailed discussion on the topic of integrity of storage sites and the approaches that can be taken for maintaining it.

The foundation of successful simultaneous CO_2-EOR and storage projects is always built on the selection of favorable storage sites. Chapter 7 briefly discusses the best practices for initial screening, selection, and initial characterization of favorable storage sites. The role of some of some simple modeling tools in performing a priori evaluations of the qualified sites that are being planned for commercial projects before embarking on time consuming and lengthy detailed characterization is also discussed. These tools can help the operators in performing initial optimization studies in making certain engineering decisions that could lead to maximum oil recovery as well as maximum storage of injected CO_2.

This book will be of interest to the workforce of the petroleum industry which is already engaged in geologic CO_2 storage projects or will be engaged in the future. For meeting the engineering challenges of designing, operating, and managing future geologic CO_2 storage projects and bringing a synergy between EOR and storage, an improved understanding of various engineering aspects of geologic CO_2 storage should be of great help to practicing engineers, managers, policy makers, and other stakeholders who are tirelessly working for increasing the acceptance of geologic CO_2 storage as an environmentally and socially sensible solution for abating the GHG emissions while establishing it as one of the core business strategies of the petroleum industry in low carbon economy.

Dr. Ramesh K. Agarwal
William Palm Professor of Engineering
Washington University in St. Louis, USA

Preface

The main motivation to write this book was to discuss various engineering aspects of geologic CO_2 storage in a manner so that the value of CO_2 injection into the subsurface as an environmentally sensible and economically profitable business activity can be appreciated by anybody interested in the topic.

Bakersfield, USA   Dayanand Saini

Acknowledgements

I would like to thank Mojtaba (Reza) Ardali and Patrick Niebuhr, who as continuing education program chair for the San Joaquin Section of the Society of Petroleum Engineers (SJV SPE) during 2015 and 2016, respectively, facilitated the offering of my one-day professional development course on geologic CO_2 storage which ultimately led me to write this book. Many thanks to several of my undergraduate and graduate research students for asking numerous thought provoking questions about the topics covered in the book.

Contents

Abbreviations

ADB	Asian Development Bank
API	American Petroleum Institute
ARI	Advanced Resources International
BPMs	Best Practice Manuals
CCS	Carbon Capture and Sequestration
CCUS	Carbon Capture, Use, and Storage
CC&ST	Carbon Capture and Sequestration Technologies
C2ES	Center for Climate and Energy Solutions
DOE	Department of Energy
DSFs	Deep Saline Formations
EPA	Environmental Protection Agency
EOR	Enhanced Oil Recovery
EOR+	Simultaneous EOR and Storage
E&P	Exploration and Production
ESI	Emirates Steel Industries
FPSO	Floating Production, Storage, and Offloading
FWU	Farnsworth Unit
GAGD	Gas Assisted Gravity Drainage
GHG	Greenhouse Gases
Gt	Gigatonnes
IEA	International Energy Agency
IEAGHG	International Energy Agency Greenhouse Gas R&D Programme
IMS	Intelligent Monitoring Networks
InSAR	Interferometric Synthetic Aperture Radar
IPCC	Intergovernmental Panel on Climate Change
K-wave	Krauklis Wave
LGT	Linear Gradient Theory
LSIPs	Large-Scale Integrated Carbon Capture and Storage Projects
MMP	Minimum Miscibility Pressure
MMV	Monitoring, Management, and Verification

MOC	Method of Characteristics
Mtpa	Million tonnes per year
MVA	Monitoring, Verification, and Accounting
NBU	North Burbank Unit
NETL	National Energy Technology Laboratory
OGCI	Oil and Gas Climate Initiative
OOIP	Original Oil in Place
OWC	Oil-Water Contact
PRC	People's Republic of China
PNC/PNL	Pulse Neutron Capture Logging
PNS	Pulse Neutron Spectroscopy
PV	Pore Volume
PVT	Pressure, Volume, Temperature
RBA	Rising Bubble Apparatus
RCSPs	Regional Carbon Sequestration Partnerships
R&D	Research and Development
RF	Recovery Factor
ROZs	Residual Oil Zones
SACROC	Scurry Area Canyon Reef Operators Committee
UAE	United Arab Emirates
UNFCC	United Nations Framework Convention on Climate Change
UNIDO	United Nations Industrial Development Organization
USDWs	Deepest Underground Sources of Drinking Water
VIT	Vanishing Interfacial Tension
VSP	Vertical Seismic Profiling
WAG	Water Alternating Gas

Chapter 1
Role of Petroleum Industry

Abstract The expertise and experience of petroleum industry around the world is playing a key role in developing and commercializing the geologic CO_2 storage, a necessity technology for achieving a balance in anthropogenic greenhouse gases (GHG) emissions by sources and removals by sinks of GHG. However, for realizing the full potential of geologic CO_2 storage as a future service industry, petroleum industry needs to overcome various obstacles and barriers including the recognition of geologic CO_2 storage as part of their core business, training of both the existing and the future workforce of the petroleum industry to prepare them for meeting the engineering challenges of designing, operating, and managing future geologic CO_2 storage projects, and an improved understanding of various engineering aspects of the simultaneous CO_2 enhanced oil recovery and storage strategy.

1.1 Introduction

Since its inception, when Colonel Edwin L. Drake struck oil in a 21 m (69 ft) deep well drilled in 1859, in Titusville, Pennsylvania, USA (Latson 2015), the modern petroleum industry has come a long way. Ingenuity, technological advancements, free enterprise, and innovation, have made it possible that oil and natural gas combinedly provide 52.5% of world's total primary energy supply (IEA 2016). At the same time, human activities and mainly the combustion of fossil fuels (coal, natural gas, and oil) for energy and transportation, certain industrial processes, and forestry and other land use (U.S. EPA 2016a), have resulted in the increase of carbon dioxide (CO_2) and other greenhouse gases (GHG) in the atmosphere. CO_2, which is the key constituent (76%) of these anthropogenic GHG emissions (U.S. EPA 2016b), is naturally present in the atmosphere as a part of the Earth's carbon cycle (the natural circulation of carbon among the atmosphere, oceans, soil, plants, and animals), and is constantly being exchanged among the atmosphere, ocean, and land surface (U.S. EPA 2016a). However, the increased level of anthropogenic GHG emissions to the atmosphere is altering the carbon cycle-both by adding more

© The Author(s) 2017

D. Saini, *Engineering Aspects of Geologic CO₂ Storage*, SpringerBriefs in Petroleum Geoscience & Engineering, DOI 10.1007/978-3-319-56074-8_1

CO_2 to the atmosphere and by influencing the ability of natural sinks, like forests, to remove CO_2 from the atmosphere (U.S. EPA 2016a).

The entering of the Paris Agreement in force on November 4, 2016 (UNFCCC 2016), in which necessity to achieve a balance in anthropogenic GHG emissions by sources and removals by sinks of GHG is clearly recognized (United Nations 2015), provides the petroleum industry an excellent opportunity to be a key player and use its expertise in CO_2-injection based enhanced oil recovery (EOR) processes in abating the GHG emissions globally via commercial-scale simultaneous CO_2-EOR and storage projects in depleted oil and gas fields and dedicated storage projects in deep saline formations (DSFs).

The underground injection and geologic sequestration (also referred to as storage) of the captured CO_2 resulting from certain human activities (new and existing coal- and gas-fired power plants and large industrial sources into deep underground rock formations (depleted oil and gas fields and DSFs) is one of the three steps in the carbon capture and sequestration (CCS) process (U.S. EPA 2016c). The other two steps involve capturing of CO_2 from power plants or industrial processes and transportation of the captured and compressed CO_2 (usually in pipelines) (U.S. EPA 2016c). Unlike terrestrial, or biologic, sequestration, where carbon is stored via agricultural and forestry practices, geologic sequestration (storage) involves injecting CO_2 deep underground where it stays permanently (U.S. EPA 2016c). There would hardly any benefit in large-scale capturing of anthropogenic CO_2, if it could not be stored permanently. Geologic CO_2 storage provides easy access to the storage sites that can store large quantities of captured CO_2 in safe manner and on permanent basis.

As per one of the projections of International Energy Agency (IEA), cutting CO_2 emissions to 5% of their 2005 levels, would require a reduction of 43 gigatonnes (Gt) of CO_2 (GtCO$_2$) and total CCS in power generation and industrial applications is expected to contribute 19% to this reduction target in 2050 (IEA 2010 in IEA/UNIDO 2011). According to another recent modeling performed by the IEA, CCS could deliver 13% of the cumulative reductions needed by 2050 (IEA 2015a).

1.2 Experience and Expertise

The petroleum industry has a long history, experience, and expertise in transporting and injecting CO_2 for effectively recovering additional oil from depleted oil fields. Much of the CO_2-injection based commercial EOR technology has been developed in North America or more precisely in the United States. Since the launching of the first commercial CO_2-EOR injection project at SACROC (Scurry Area Canyon Reef Operators Committee) Unit of the Kelly-Snyder Field in Scurry County, West Texas in 1972 (Meyer 2007), nearly 1 Gt (gigatonne) of CO_2 has been injected into geological reservoirs as part of CO_2-EOR activities worldwide (Global CCS Institute 2013). Almost all the CO_2-EOR projects are in the US with very few projects in other countries including Canada, Turkey, Abu Dhabi, China, Malaysia, and Brazil. Miscible CO_2-EOR, in which injected CO_2 and reservoir oil mix together in all proportions to form a single phase, dominates the gas injection based EOR projects.

It is estimated that in the US, today, a total of 136 CO_2-EOR projects inject 68 million tonnes per year of natural and industrial CO_2 for producing 300,000 barrels per day (b/d) of oil via EOR (Wallace and Kuuskraa 2014). However, 80% of the CO_2 used in the US CO_2-EOR operations is still derived from naturally occurring CO_2 deposits, which are limited. However, in their CO_2 supply outlook, Wallace and Kuuskraa (2014) project that the share of industrial CO_2 for in the US CO_2-EOR projects is expected to grow from current 20% to around 48% by 2020.

In CO_2-EOR operations, a significant portion of injected CO_2 is lost in the reservoir anyway. It is estimated that CO_2-EOR projects in the US stored nearly 14 million tonnes of industrial (natural gas processing and other industrial processes) CO_2 in underground formations in 2014 (Wallace and Kuuskraa 2014). Generally, CO_2-EOR operations have not been designed with long-term CO_2 storage in mind. To qualify a CO_2-EOR injection site as a storage site also, additional storage-focused activities, frequently referred as monitoring, verification, and accounting (MVA) or monitoring, management, and verification (MMV) programs to be undertaken before, during and following CO_2-injection for demonstrating and ensuring long-term (~ 1000 years) storage of 99% of injected CO_2. The Intergovernmental Panel on Climate Change (IPCC) targets that a well selected site should retain 99% CO_2 in the reservoir over 1000 years (IPCC 2007). However, CO_2-EOR operators and/or regulatory authorities may not operate or permit the injection sites for GHG mitigation purposes (Global CCS Institute 2016a) as MVA/MMV programs represent an additional cost to CO_2-EOR operators that would, if not offset by compensatory measures or revenue, negatively impact project economics (IEA 2015b).

The 1973–74 Arab oil embargo had brought the US "energy crunch" to the "front and center" and it led the petroleum industry to develop the commercial CO_2-EOR technology for reducing the dependence on foreign oil by increasing domestic production. Similarly, an urgency to combat the global GHG emissions has brought a new business opportunity to the petroleum industry. It is in the form of geologic CO_2 storage by playing a key role in achieving a balance between anthropogenic GHG emissions by sources and removal by sinks, thus, remaining competitive in an eventual low carbon economy.

The successful development and commercialization of CO_2-EOR technology was the result of the U.S. Department of Energy's (DOE's) efforts when the DOE launched a high-risk cost-shared EOR technology field development program. Since then, the US CO_2-EOR industry has emerged as a self-sustained business. Currently, numerous petroleum exploration & production (E&P) companies in collaboration with transportation, and entities with source of industrial CO_2 are not only successful in recovering additional oil in cost effective manner, but they are also on their way to store 1–2 billion tonnes of industrial CO_2 that would otherwise be vented during the next 20 years (Wallace and Kuuskraa 2014).

1.3   Current Role

Most of the CO_2 injection aspects into the reservoirs for EOR have been used in engineering practice for more than 40 years, but, CO_2 injection for geologic CO_2 storage is relatively new. Also, CO_2 injection for EOR and for geologic storage is carried out for achieving different objectives. In EOR, maximum oil recovery with minimum consumption of injected CO_2 is desired, whereas, storage projects look for achieving maximum and safe storage of injected CO_2.

The implementation of CCS technology via simultaneous EOR and storage projects is an attempt to achieve a synergy between these two but entirely different business objectives. Like the success story of commercial CO_2-EOR technology, petroleum industry has embraced the CCS technology that is primarily built on the concept of geologic CO_2 storage. In last two decades, a significant growth in the portfolio of large-scale integrated carbon capture and storage projects (LSIPs) has been observed. Projects categorized by the Global CCS institute as LSIPs must inject anthropogenic CO_2 into either dedicated geological storage sites and/or CO_2-EOR operations in depleted oil fields. Per the Global CCS Institute's online database (Global CCS Institute 2016b), globally, there are 38 large-scale integrated carbon capture and storage projects (LSIPs). Out of 38 LSIPs, 15 LSIPs are in operation (Table 1.1), with further 7 LSIPs under construction. Rest are in planning (identify, evaluate, define) or in execution stages before becoming operational. The 22 LSIPs (operational or under construction) represent a doubling since the start of this decade. The total CO_2 capture capacity of these 22 projects is around 40 million tonnes per annum (Mtpa). Apart from the LSIPs, hundreds of pilot and demonstration-scale projects and CCS initiatives that have made, or are making, a significant contribution to the understanding of CCS technologies (Global CCS Institute 2016c).

As can be seen from Table 1.1, 11 LSIPs are simultaneous CO_2-EOR and storage projects (depleted oil fields), whereas, 4 LSIPs are dedicated storage projects (DSFs). Interestingly, the US has 7 operational LSIPs and all of them are simultaneous CO_2-EOR and storage projects with a CO_2 capture capacity of around 20 Mtpa. Again, it could not be possible without the active partnership of the petroleum industry with the governmental agencies like the U.S. DOE. In the U.S., DOE's Regional Carbon Sequestration Partnerships (RCSPs) initiative, launched in 2003, forms the centerpiece of national efforts to develop the infrastructure and knowledge base needed to place carbon storage and utilization technologies on the path to commercialization (U.S. DOE 2016). The seven RCSPs are comprised of more than 400 diverse organizations covering 43 states and four Canadian provinces and currently. Currently, the RCSPs, in collaboration with CO_2-EOR operating companies like Denbury Resources, Cenovus Energy, Apache Canada, Chaparral Energy, and Merit Energy are working to demonstrate the long-term, effective, and safe storage and utilization of CO_2 in the USA and Canadian depleted oil fields.

Apart from the USA and Canada, active industry-government collaboration has also resulted in successful implementation of commercial CCS technology

Table 1.1 Operational large-scale integrated carbon capture and storage projects (LSIPs) globally (Global CCS Institute 2016b)

Project name	CO$_2$ capturing facility location	CO$_2$ capturing operation Date	Source of anthropogenic CO$_2$ (industry)	Capture type	Current capture capacity (million tonnes per year)	Transport type	Primary storage type	Major storage site (depleted oilfield)	Storage site location	CO$_2$ injection start date (simultaneous CO-EOR and storage project)	Current CO$_2$ injection rate (million tonnes per year)	CO$_2$ stored to date (million tonnes)
Great Plains Synfuel Plant	North Dakota, USA	2000	Synthetic natural gas	Pre-combustion capture (gasification)	3.0	Pipeline	EOR	Weyburn and Midale Oil Units	Saskatchewan, Canada	2000	2.4 (Weyburn) 0.6 (Midale)	19.2 (Weyburn, Dec 2012) 2.75 (Midale, 2012) (NETL 2015a)
Boundary Dam Carbon Capture and Storage Project	Saskatchewan, Canada	2014	Power generation (Coal-fired)	Post-combustion capture	1.0	Pipeline	EOR	Weyburn Oil Unit	Saskatchewan, Canada	–	–	–
Lost Cabin Gas Plant	Wyoming, USA	2013	Natural gas processing	Pre-combustion capture (natural gas processing)	0.9	Pipeline	EOR	Bell Creek	Montana, USA	2013	>0.9	2.75 (May 2016) (Albenze 2016)
Val Verde natural Gas Plants	Texas, USA	1972	Natural gas processing	Pre-combustion capture (natural gas processing)	1.3	Pipeline	EOR	SACROC Unit (Kelly Snyder oilfield)	Texas, USA	2008	–	55.00 (since 1972) (Grigg et al. 2012)
Air Products Steam Methane Reformer EOR Project	Texas, USA	2013	Hydrogen production	Industrial separation	1.0	Pipeline	EOR	West Hastings Oilfield	Texas, USA	2010	1.0	3.00 (May, 2016) (NETL 2015b)

(continued)

Table 1.1 (continued)

Project name	CO_2 capturing facility location	CO_2 capturing operation Date	Source of anthropogenic CO_2 (industry)	Capture type	Current capture capacity (million tonnes per year)	Transport type	Primary storage type	Major storage site (depleted oilfield)	Storage site location	CO_2 injection start date (simultaneous CO-EOR and storage project)	Current CO_2 injection rate (million tonnes per year)	CO_2 stored to date (million tonnes)
Shute Creek Gas Processing Facility	Wyoming, USA	1986	Natural gas processing	Pre-combustion capture (natural gas processing)	7.0	Pipeline	EOR	Number of depleted oilfields in Wyoming and Colorado (also supplies CO_2 to Bell Creek Oilfield)	Wyoming, Colorado, Montana, USA	–	–	–
Enid Fertilizer CO_2-EOR Project	Oklahoma, USA	1982	Fertilizer production	Industrial separation	0.7	Pipeline	EOR	Several depleted oilfields south of Oklahoma City	Oklahoma, USA	–	–	–
Coffeyville Gasification Plant	Kansas, USA	2013	Fertilizer production	Industrial separation	1.0	Pipeline	EOR	North Burbank Oil Unit	Oklahoma, USA	–	–	–
Century Plant	Texas, USA	2010	Natural gas processing	Pre-combustion capture (natural gas processing)	8.4	Pipeline	EOR	Many oilfields in the Permian basin	Texas, USA	–	–	–
Michigan Basin Project	Michigan, USA	2013	Shale gas processing	Pre-combustion capture (natural gas processing)	–	Pipeline	EOR	Several pinnacle reefs within Michigan's Northern Reef Trend	Michigan, USA	2013	0.4	0.49 (May 2016) (Albenze 2016)
Farnsworth Unit Project	Kansas, USA (Arkalon Ethanol Plant), Texas, USA (Agrium Fertilizer Plant)	–	Ethanol and Fertilizer production	Industrial separation	–	Pipeline	EOR	Farnsworth Unit	Texas, USA	2010	0.2	0.42 (May 2016) (Albenze 2016)

involving geologic CO_2 storage as evident from operational LSIPs in Norway, Brazil, Saudi Arabia, and Algeria. Other countries including China, Australia, South Korea, Netherlands, and the United Arab Emirates are also witnessing the successful implementation of LSIPs.

As Lovell (2010) nicely points out the relevance of geologic CO_2 storage to the petroleum industry. *"Pumping waste into long-term storage is not what we veteran frontier explorers are used to, with our techno-gamble culture of high risk and high reward."* He further asserts that geologic CO_2 storage is a future service industry, with a price per tonnes for all carbon safely stored. Why should the petroleum industry not seize this opportunity?

1.4   Future Challenges

Before, the petroleum industry realizes the full potential of geologic CO_2 storage, it needs to overcome various obstacles and barriers. Currently, main source of CO_2 supply to operational LSIPs across the globe is pre-combustion capture (natural gas processing) and industrial separation. CO_2 capturing at coal-fired power plants, which are the top source of CO_2 emissions, is still a distant goal.

It is often expensive to retrofit on existing coal-fired power plants and utilities companies in countries like the USA and Canada may have little interest in doing so as the role of coal-fired plants for electricity generation is declining in favor of natural gas and other energy sources due to low natural gas prices, state renewable energy standards and environmental regulations (C2ES 2016). Another key barrier is the lack of private-public partnership in mostly-nationalized petroleum (exploration & production) industry in countries like China and India. Hence, making the operational LSIPs a reality in China and India will take more time compared to it took in countries like the US, Canada, where, private operators had significant experience and expertise in injecting CO_2 for EOR purpose.

Lack of rational economic reason for including geological CO_2 storage in core business activities of petroleum E&P companies is another significant barrier. Though, increasingly petroleum E&P companies are, now, recognizing and realizing the promise that geologic CO_2 storage may hold for their core business, however, they, still, seem to hesitant in spending investor's money on it. Yes, injecting CO_2 for recovering additional oil makes a perfect business case, however, lack of any substantial return on the investment in case of dedicated storage projects and the additional cost associated with mandatory MVA/MMV program needed for ensuring permanent retention of injected CO_2 in injection zone over long period (~ 1000 years) of time, often, prevent the small and independent petroleum companies from launching geologic CO_2 storage projects.

Lack of strong incentives and policy and economic drivers that could spur the necessary innovations and lead to the development of cost effective and efficient, injection, storage, and monitoring technologies for creating a competitive market for carbon storage is also a big challenge in the way of making geologic CO_2 storage a service industry of the future.

The training of both the existing and the future workforce of the petroleum industry to prepare them for meeting the engineering challenges of designing, operating, and managing future geologic CO_2 storage projects is also going to be a key challenge. It is estimated that up to 900 Gt CO_2 could be eventually stored in depleted oil reservoirs worldwide (IPCC 2007). Conversion of this theoretical storage capacity into "realistic and commercially achievable" storage capacity could not be possible without a well-trained engineering workforce. The education of stakeholders and policy makers for better understanding of commercial, cost, regulatory, and social barriers will also be one of the key challenges that petroleum industry needs to overcome for creating an economic value for the industry via geologic CO_2 storage.

The potential negative environmental impacts such as induced seismicity, possible ground water contamination due to above injection zone migration, and worst case scenario of the leakage of injected CO_2 to the surface and/or back into the atmosphere, are other key challenges that petroleum industry will need to face if geologic CO_2 storage projects are implemented at wider scale in future. Such engineering challenges would not be addressed without inclusion of appropriate commercially available monitoring tools and technologies.

Nevertheless, a synergy between CO_2-EOR and storage that has already been demonstrated by the public-private (governmental agencies, academia and research organizations, and petroleum industry) by successfully designing, operating, and managing currently operational LSIPs and other large-scale simultaneous CO_2-EOR and storage projects can be the key of wide-scale commercial implementation of geologic storage projects.

An improved understanding of various engineering aspects of simultaneous CO_2-EOR and storage strategy such as MVA/MMV program, injection strategies, CO_2-oil miscibility, integrity of injection zone(s) and overlying seal(s), advances in screening of suitable candidate oil fields, well configuration schemes, and other engineering practices for maximizing oil recovery and storage capacity will definite help the petroleum industry (practicing and aspiring engineers, managers) and community (policy makers, stakeholders) in increasing the acceptance of geologic CO_2 storage as one of the most commercially lucrative and environmentally and socially sensible solutions for abating the GHG emissions.

References

Albenze E (2016) NETL regional partnerships-overview. http://www.sequestration.org/resources/PAGMay2016Presentations/02-Albenze_RCSP_Overview_2016.pdf. Accessed 25 Dec 2016
C2ES (2016) Coal. Center for Climate and Energy Solutions. https://www.c2es.org/energy/source/coal. Accessed 22 Jan 2017
Global CCS Institute (2013) The global status of CCS: 2013. Melbourne, Australia

Global CCS Institute (2016a) Large-scale CCS projects—definitions. http://www.global ccsinstitute.com/projects/large-scale-ccs-projects-definitions. Accessed 22 Jan 2017

Global CCS Institute (2016b) Large-scale CCS projects (project database). http://www. globalccsinstitute.com/projects/large-scale-ccs-projects#map. Accessed 22 Jan 2017

Global CCS Institute (2016c) Pilot and demonstration projects (Project database). https://www. globalccsinstitute.com/projects/pilot-and-demonstration-projects. Accessed 14 Mar 2017

Grigg RB, McPherson B, Lee R (2012) Phase II final scientific/technical report. Southwest Regional Partnership on Carbon Sequestration. New Mexico Institute of Mining and Technology, Socorro, New Mexico. Available via https://www.southwestcarbonpartnership.org/download/the-swp/phase-ii/PhaseII_Final_SWP_2.pdf. Accessed 24 Dec 2016

IEA/UNIDO (2011) Technology roadmap (carbon capture and storage in industrial applications). International Energy Agency and The United Nations Industrial Development Organization. Available via https://www.unido.org/fileadmin/user_media/Services/Energy_and_Climate_Change/Energy_Efficiency/CCS/CCS_Industry_Roadmap_WEB.pdf. Accessed 22 Jan 2017

IEA (2015a) Carbon capture and storage: the solution for deep emissions reductions. International Energy Agency. Available via https://www.iea.org/publications/freepublications/publication/CarbonCaptureandStorageThesolutionfordeepemissionsreductions.pdf. Accessed 22 Jan 2017

IEA (2015b) Storing CO_2 through enhanced oil recovery (combining EOR with CO_2 storage (EOR+) for profit. International Energy Agency. Available via https://www.iea.org/publications/insights/insightpublications/Storing_CO2_through_Enhanced_Oil_Recovery.pdf. Accessed 22 Jan 2017

IEA (2016) Key world energy statistics. International Energy Agency. https://www.iea.org/publications/freepublications/publication/KeyWorld2016.pdf. Accessed 22 Jan 2017

IPCC (2007) Climate change 2007: Impacts, adaptation, and vulnerability. In: Parry ML el al (eds) Contribution of working Group II to the fourth assessment report of the Intergovernmental Panel on Climate Change, Cambridge University Press, Cambridge, UK, pp 976

Latson J (2015) How the American oil industry got its start. http://time.com/4008544/american-oil-well-history/. Accessed 20 Dec 2016

Lovell B (2010) Challenged by carbon (the oil industry and climate change). Cambridge University Press, Cambridge, UK

Meyer JP (2007) Summary of carbon dioxide enhanced oil recovery (CO2EOR) injection well technology. American Petroleum Institute. Available via http://www.api.org/~/media/files/ehs/climate-change/summary-carbon-dioxide-enhanced-oil-recovery-well-tech.pdf. Accessed 22 Jan 2017

NETL (2015a) Project facts, IEA-GHG Weyburn-Midale CO_2 monitoring and storage project. The U.S. Department of Energy's (DOE) National Energy Technology Laboratory. https://www.netl.doe.gov/publications/factsheets/project/Proj282.pdf. Accessed 25 Dec 2016

NETL (2015b) News release (Texas CO_2 capture demonstration project hits two million metric ton milestone). The U.S. Department of Energy's (DOE) National Energy Technology Laboratory. https://netl.doe.gov/newsroom/news-releases/news-details?id=f63a9d2e-6936-4c17-913c-1c59 5864978e. Accessed 25 Dec 2016

U.S. DOE (2016) Regional partnerships. U.S. Department of Energy (Office of Fossil Energy). https://energy.gov/fe/science-innovation/carbon-capture-and-storage-research/regional-partner ships. Accessed 20 Dec 2016

UNFCCC (2016) The Paris agreement. United Nations Framework Convention on Climate Change. http://unfccc.int/paris_agreement/items/9485.php. Accessed 20 Dec 2016

United Nations (2015) Paris agreement. http://unfccc.int/files/essential_background/convention/application/pdf/english_paris_agreement.pdf. Accessed 20 Dec 2016

U.S. EPA (2016a) Carbon dioxide emissions. In: Overview of greenhouse gases. U.S. Environmental Protection Agency. https://www.epa.gov/ghgemissions/overview-greenhouse-gases#carbon-dioxide. Accessed 20 Dec 2016

U.S. EPA (2016b) Global emissions by gas. In: global greenhouse gas emissions data. U.S. Environmental Protection Agency. https://www.epa.gov/ghgemissions/global-greenhouse-gas-emissions-data. Accessed 20 Dec 2016

U.S. EPA (2016c) Carbon dioxide capture and sequestration: overview. U.S. Environmental protection agency. https://www.epa.gov/climatechange/carbon-dioxide-capture-and-sequestration-overview#CCS. Accessed 20 Dec 2016

Wallace M, Kuuskraa V (2014) Near-Term projections of CO_2 utilization for enhanced oil recovery. U.S. Department of Energy, National Energy Technology Laboratory. Report no. DOE/NETL-2014/1648

Chapter 2
Simultaneous CO$_2$-EOR and Storage Projects

Abstract Majority of currently operational large-scale integrated carbon capture and storage projects (LSIPs) are in the USA and Canada. This demonstrates the leadership of the North American petroleum industry in successful implementation of geologic CO$_2$ storage technology via simultaneous CO$_2$-EOR and storage strategy. These LSIPs mainly utilize anthropogenic CO$_2$ captured at the natural gas processing plant used by the petroleum industry for processing the natural gas resulting from its routine oil and gas extraction operations. Other oil producing countries such Saudi Arabia, United Arab Emirates (UAE), and Brazil are also following suite by launching simultaneous CO$_2$-EOR and storage strategy based LSIPs and large-scale geologic CO$_2$ storage projects.

2.1 Introduction

For more than four decades, the petroleum industry in the USA has been using the anthropogenic CO$_2$ for EOR purposes. In CO$_2$-EOR operations, a significant portion of injected CO$_2$ is lost in the reservoir in anyway leading to its partial (incidental) storage even though they are not designed with long-term storage purposes. With an inclusion of additional storage-focused activities (i.e. a dedicated MVA/MMA program) a EOR project can become a storage project (i.e. simultaneous CO$_2$-EOR and storage project). A MVA/MMA program essentially includes a minimum of following four activities (IEA 2015b):

1. additional site characterization and risk assessment to evaluate the storage capability of a site;
2. additional monitoring of vented and fugitive emissions;
3. additional subsurface monitoring, and
4. change to field abandonment practices.

The petroleum industry's long and successful record of secure underground injection of CO$_2$ for EOR, has helped the world to embrace the geologic CO$_2$ storage as first-order technology for abating the anthropogenic GHG emissions. It is

© The Author(s) 2017   11
D. Saini, *Engineering Aspects of Geologic CO$_2$ Storage*, SpringerBriefs
in Petroleum Geoscience & Engineering, DOI 10.1007/978-3-319-56074-8_2

substantiated from the fact that in 12 out of 15 currently operational LSIPs, primary storage type is EOR. The Global CCS Institute (2016a) defines the LSIPs as projects involving the capture, transport, and storage of CO$_2$ at a scale of:

1. at least 800,000 metric tons (tonnes) of CO$_2$ annually for a coal–based power plant, or
2. at least 400,000 metric tons (tonnes) of CO$_2$ annually for other emissions–intensive industrial facilities (including natural gas–based power generation).

Projects categorized by the Global CCS institute as LSIPs must inject anthropogenic CO$_2$ into either dedicated geological storage sites and/or enhanced oil recovery (CO$_2$-EOR) operations. Obviously, majority (9) of LSIPs are in North America [the USA (7) and Canada (2)] where the petroleum industry has already mastered the commercial CO$_2$-EOR technology. Brazil, Saudi Arabia, and the United Arab Emirates, each, have one simultaneous EOR and storage LSIP.

Apart from LSIPs, numerous pilot and demonstration projects and commercial CO$_2$-EOR projects which use either natural or anthropogenic CO$_2$, have also helped in gaining necessary information on technical feasibility of various capture, injection, storage, and monitoring technologies and in gaining operational experience within the scope of policy, regulatory, and project economics framework.

2.2 North-American Projects

If we look at the project histories of the current large-scale North American simultaneous EOR and storage projects (Table 1.1), The Great Plains Synfuel Plant LSIP in southwestern North Dakota is the only commercial-scale coal gasification plant in the US and has been capturing and transporting CO$_2$ to oil fields in Canada since October 2000 (Global CCS Institute 2016d). The captured CO$_2$ is transported via (the Souris Valley) pipeline to the Weyburn and Midale Oil Units in Saskatchewan, Canada. The International Energy Agency (IEA) as a part of its Greenhouse Gas R&D Programme (IEAGHG), performed the most comprehensive MVA/MMA program alongside CO$_2$-EOR operations between 2000 and 2011. The IEAGHG Weyburn-Midale CO$_2$ Monitoring and Storage Project, as it is called, is the largest full-scale CCS field study ever conducted that included the study of mile-deep seals that contain the CO$_2$ reservoir, CO$_2$ plume movement, and the monitoring of permanent storage (Global CCS Institute 2016d). In October 2014, the Weyburn Oil Unit also started to receive the CO$_2$ captured at the Boundary Dam Carbon Capture and Storage Project LSIP (Global CCS Institute 2016e).

CO$_2$ injection in the Vale Verde Natural Gas Plants LSIP in Texas has ongoing since 1972. However, at the SACROC Unit of the Kelly Snyder oil field which is the largest storage site among many nearby sites, new injection coupled with dedicated MVA/MMV commenced in 2008 (Grigg et al. 2012). In case of another LSIP in Texas, namely, the Air Products Steam Methane Reformer EOR Project,

CO_2 injection in the historic West Hastings oil field began in 2010 and a research MVA program to study the movement and sequestration of CO_2 through existing EOR operations was implemented and is continuing. Both the Lost Cabin Gas Plant and the Shut Creek Gas Processing Facility in Wyoming, provide the CO_2 for the Denbury Resources-operated Bell Creek oil field in Montana (NETL 2015c). More than 230 miles long Greencore pipeline supplies the CO_2 for the Bell Creek site from these LSIPs. The Bell Creek Project had collected a relevant baseline MVA data before injection began in May 2013 and a continued robust MVA program is in place as project is moving forward with its injection, production, and storage operations.

Two other LSIPs, namely, Enid Fertilizer CO_2-EOR Project and Shut Creek Gas Processing Facility are operational for more than three decades and supply anthropogenic CO_2 to several depleted oil fields in Oklahoma and Wyoming, respectively. As mentioned above, the Shut Creek LSIP also supplies CO_2 to the Bell Creek site. Similarly, captured CO_2 at Century Plant in Texas is distributed to many oil fields in the Permian basin. The Coffeyville Gasification Plant LSIP in Kansas is the source of CO_2 for the North Burbank Unit (NBU) which is the one of the largest oil fields in Oklahoma and was originally discovered in 1920. CO_2 injection into NBU for simultaneous EOR and storage purposes started in June 2013. However, for various reasons including injection and storage of supplied CO_2 into multiple sites, there is little information on the MVA/MMV programs for these three LSIPs is available in public domain. Nevertheless, these LSIPs are playing a key role in establishing simultaneous CO_2-EOR and storage strategy as a commercially viable option for geologic CO_2 storage.

Apart from the operational LSIPs, there are two notable simultaneous CO_2-EOR and storage projects in North America, namely, Michigan Basin Project, Michigan, and Farnsworth Unit Project, Texas. In Michigan Basin Project, CO_2 injection into an oil field located within Michigan's Northern Reef Trend started in April 2013 and the monitoring and tracking of the injected CO_2 was started in July 2013 (MIT CC&ST Program 2016). The Chaparral Energy began CO_2 injection in December 2010 into the Morrow Sandstone Formation within the Farnsworth Unit (FWU) for EOR and the MVA program was launched in October 2013.

2.2.1 Key Features

Table 2.1 along with Table 1.1 provide a summary of key geologic characteristics, reservoir parameters, and other operational characteristics and statistics of the main storage sites of simultaneous CO_2-EOR and storage strategy based LSIPs and other large-scale projects that are currently operational in the North America. It is remarkable that majority of these projects (9 out of 11) rely on the CO_2 captured by natural gas processing or industrial separation units (Table 1.1). The natural gas

Table 2.1 Key geologic characteristics and reservoir parameters of the storage sites of some of the currently operational North American LSIPs and other large-scale simultaneous CO_2-EOR and storage projects

Geologic characteristic/reservoir parameter[a]	Unit	Weyburn oil unit	Bell Creek	SACROC unit	West hastings	North Burbank oil unit	Pinnacle Reefs (Michigan's Northern Reef Trend)	Farnsworth unit
Formation		Charles formation [Marly (upper dolostone unit) + Vuggy (lower limestone unit)]	Muddy (Newcastle)	Canyon Reef (limestone)	Frio sandstone	Burbank Sandstone	Guelph formation (brown Niagaran)	Upper morrow
Geological age		Mississippian	Cretaceous	Pennsylvanian	Oligocene	Pennsylvanian	Silurian	Pennsylvanian
Hydrocarbon trap type		Truncated stratigraphic	Stratigraphic	Reef	Structural	Stratigraphic	Reef	Stratigraphic
Overlying caprock(s)		Midale evaporite with Watrous aquitard as regional seal	Mowry shale	Wolfcamp shale	Anahuac shale	Cherokee shale	A-2 evaporite (top) A-1 evaporite (flank)	Thirteen Finger limestone
Caprock(s) average thickness	ft.	6.5–36 (Midale evaporite)	>3000	600–1100	600	45–70	>290	118
Formation depth	ft.	4900	4500	6200–7000	5500	3000	5400–5700	7545–7950
Avg. reservoir thickness	ft.	19.5 (Marly) 49 (Vuggy)	30–45	229	>700	50	278 (maximum)	54
Formation pressure at discovery	psi	2300	1180	3122–3300	2740	1350–1600	2400	2200
Formation temperature	°F	138	110	130	160	122	108	167
Cumulative oil production to date	Million barrels	366	133	1400	582		0.47 (Dover 33)	19
Oil gravity	°API	25–34	32–41	42	31	39–41	47.9	38
Formation water salinity	ppm	20,000–310,000	5000	159,000	>100,000	85,000	Very high	3600

(continued)

Table 2.1 (continued)

Geologic characteristic/reservoir parameter[a]	Unit	Weyburn oil unit	Bell Creek	SACROC unit	West hastings	North Burbank oil unit	Pinnacle Reefs (Michigan's Northern Reef Trend)	Farnsworth unit
Avg. porosity	%	26 (Marly) 11 (Vuggy)	25–35	9	29	20	4	3–21
Avg. permeability	mD	10 (Marly) 15 (Vuggy)	150–1175	30	500–1000	50–80	12	0.1–700
EOR type		Combined miscible simultaneous but separate CO_2 only, water only, and water alternating gas injection strategy using a combination of horizontal CO_2 injectors and horizontal producers and vertical water injectors and vertical producers	Continuous miscible CO_2 injection (5-spot pattern)	Miscible Water Alternating Gas (WAG) (5-spot well pattern)	Continuous miscible CO_2 injection, water only, and water alternating gas (5-spot pattern)	Miscible Water Alternating Gas (WAG) (staggered line drive well pattern)	Top down CO_2 injection (vertical injector + horizontal producer)	Hybrid water alternating with CO_2 gas injection (WAG) (5-spot well pattern)
Reported reservoir pressure prior to CO_2 injection	psi	2150–2250	1572	2400	1800	900	790	4700

[a]Data sources include Ahmmed (2015), Ampomah et al. (2016), (Balch and McPherson 2016), Davis et al. (2014), Ganesh et al. (2014), Global CCS Institute (2016d), Gorecki et al. (2014), Grigg et al. (2012), Han (2010), Meng (2015), Li and Schechter (2014), Miller et al. (2014), MIT CC&ST Program (2016), NETL (2015c), Pan et al. (2016), Riding and Rochelle (2005), White et al. (2014), Whittaker (2005), Whittaker et al. (2011), Wood et al. (2006)

processing facilities are the one of the essential elements of the petroleum industry's core business i.e. oil and gas production operations. Only the Boundary Dam and the Great Plains Synfuel are the two coal-based (power generation or synthetic natural gas production) facilities that supply the Weyburn-Midale CO$_2$-EOR and storage project in Canada.

Most of the future LSIPs around the world (Global CCS Institute 2016d) will also rely on petroleum industry's ability of capturing CO$_2$ at its natural gas processing facilities. However, there appears to be a major push by China to capture CO$_2$ at its coal-fired power plants and use it for simultaneous CO$_2$-EOR and storage projects. If successful, it is going to open the door for the future carbon storage industry that is mostly absent in countries like China and India, where coal-fired plants will still be the main source of electricity generation for several decades to come (EIA 2016).

It is worth to mention here that all the storage sites (Table 2.1) had significant history of waterflooding and/or CO$_2$ injection prior to launch of MVA/MMV programs there. It implies that petroleum industry already had the infrastructure and technical knowhow needed for injecting large quantities of fluids (water or CO$_2$) into porous media. Also, the industry relies on a mix of traditional water alternating gas (WAG) and a top down continuous CO$_2$ injection in near miscible/miscible mode strategy for recovering additional oil and storing large quantities of CO$_2$ in these storage sites (depleted oil fields). Vertical wells are being used for injection and production is all but two operational projects. The Michigan Basin and the Weyburn-Midale Projects, horizontal wells have also been used for production and both injection and production, respectively.

Because, CO$_2$ injection is going on in SACROC Oil Unit (storage site for Val Verde LSIP) since 1972, it is obvious that it has stored the maximum CO$_2$ (55 million tonnes) among all operational large-scale North American simultaneous CO$_2$-EOR and storage project. The Weybun-Midale LSIP is operational since 2000 and have stored almost 22 million tonnes of CO$_2$ so far. In both the West Hastings (site for Air Products LSIP) and the Bell Creek (Lost Cabin LSIP) oil fields, CO$_2$ injection was started in 2013 and, by May 2016, both have stored 3 million tonnes and 2.75 million tonnes of captured CO$_2$, respectively.

Interestingly, majority of the storage sites are either stratigraphic traps or closed pinnacle reef structures encased in thick impermeable formations that have served as effective seals for the hydrocarbon accumulations at first place. Obviously, significant production (cumulative oil production) and injection (water and/or CO$_2$) histories of these depleted oil fields indicate that these sites can store large amounts of injected fluids. It has given additional confidence to the operators to select these sites as prime locations for storing anthropogenic CO$_2$. The access to wealth of geologic and reservoir characterization data resulting from the industry's efforts to recover stranded oil from these depleted oil fields appear to be another great reason to select them as first-order storage sites.

2.3   Current Projects (Rest of the World)

2.3.1   Uthmaniyah CO_2-EOR Demonstration Project

Even though the Kingdom of Saudi Arabia has abundant conventional hydrocarbon reserves and EOR is not likely to be required at production scale for decades to come, Uthmaniyah CO_2-EOR Demonstration Project is meant to demonstrate the proactive approach of Saudi Aramco, and industry leader and operator of the project, for addressing global environmental challenges. CO_2 at the injection site (a small flooded area in the Uthmaniyah production unit) comes from the Hawiyah Natural Gas Liquids (NGL) Recovery Plant via an 85 km (52 miles) long pipeline and is injected into Jurassic organic-rich mudstones at a depth of between 1800 and 2100 m (6000–7000 ft.) at a rate around 0.8 million tonnes per year (Global CCS Institute 2016f). The injection site includes four injection wells, four producers, and two observation wells. Injection strategy is conventional WAG. The project design is based on reservoir simulation studies and has a comprehensive monitoring and surveillance plan, including routine and advanced logging and use of new technologies for plume tracking and for CO_2 saturation modeling (seismic, chemical tracers, and electromagnetic surveys and borehole gravimetry) (Global CCS Institute 2016f).

2.3.2   Abu Dhabi CCS Project (Phase 1: ESI CCS Project)

In November, 2016, world's first commercial carbon capture facility at Emirates Steel Industries (ESI) steel production plant in Abu Dhabi, United Arab Emirates (UAE), started to capture around 0.8 million tonnes CO_2 per year to supply it via a 43 km (27 miles) pipeline for EOR injection into NEB (Al Rumaitha) and Bab onshore oil fields of the Abu Dhabi National Oil Company (ADNOC) (Global CCS Institute 2016g). Prior to launching the project, operators undertook a pilot project that involved injection of approximately 60 tonnes of CO_2 per day into the ADNOC Al Rumaitha oilfield via a CO_2 injection well Global CCS Institute 2016g). The pilot project also included an observation well and an oil producing well. The pilot provided information on the amounts of CO_2 required for field-scale injection and the potential volume of oil recoverable from the Al Rumaitha and Bab oil fields.

2.3.3   Petrobras Santos Basin Pre-salt Oil Field CCS Project

This Brazilian simultaneous CO_2-EOR and storage project is an offshore (Santos Basin) project in which pre-combustion capturing (natural gas processing) of CO_2 is done at floating production, storage, and offloading (FPSO) vessels anchored in the Santos Basin. The captured CO$_2$ is injected at a rate of approximately 1 million tonnes per year into the pre-salt carbonate reservoir of the Lula and Sapinhoá oil fields at a depth of between 5000 and 7000 m (16,400–23,000 ft.) below sea level (Global CCS Institute 2016h).

Because both Uthmaniyah and Abu Dhabi projects have become operational recently, lesson learned in these projects will take time to become available in public domain. Being an offshore project, the Petrobras CCS Project, is a unique project and lesson learned there will provide valuable insights for the future geologic CO$_2$ storage projects in offshore environment.

References

Ahmmed B (2015) Numerical modeling of CO$_2$-water-rock interactions in the Farnsworth, Texas hydrocarbon unit, USA. MS Thesis. The University of Missouri, Columbia. Available via https://mospace.umsystem.edu/xmlui/bitstream/handle/10355/46985/research.pdf?sequence=2&isAllowed=y. Accessed 23 Jan 2017

Ampomah W, Balch RS, Grigg RB et al (2016) Farnsworth Field CO$_2$-EOR project: performance case history. Soc Pet Eng. doi:10.2118/179528-MS

Balch R, McPherson B (2016) Integrating enhanced oil recovery and carbon capture and storage projects: a case study at farnsworth field. Soc Pet Eng, Texas. doi:10.2118/180408-MS

Davis D et al (2011) Large scale CO$_2$ flood begins along Texas Gulf coast (technical challenges in re-activating an old oil field). 17th annual CO$_2$ flooding conference, Midland, Texas. Available via http://www.co2conference.net/wp-content/uploads/2012/12/3.2-Denbury_Davis_Hastings_2011-CO2Flooding_Conf.pdf. Accessed 24 Dec 2016

EIA (2016) International Energy Outlook 2016 (with projections to 2040). U.S. Energy Information Administration. Available via https://www.eia.gov/outlooks/ieo/pdf/0484(2016).pdf. Accessed 14 Mar 2017

Ganesh PR et al (2014) Assessment of CO$_2$ injectivity and storage capacity in a depleted pinnacle reef oil field in northern Michigan. Energy Procedia 63:2969–2976

Global CCS Institute (2016a) Large-Scale CCS projects—definitions. http://www.globalccsinstitute.com/projects/large-scale-ccs-projects-definitions. Accessed 22 Jan 2017

Global CCS Institute (2016d) Great Plains synfuel plant and Weyburn-Midale project. http://www.globalccsinstitute.com/projects/great-plains-synfuel-plant-and-weyburn-midale-project. Accessed 25 Dec 2016

Global CCS Institute (2016e) Boundary Dam carbon capture and storage project. http://www.globalccsinstitute.com/projects/boundary-dam-carbon-capture-and-storage-project. Accessed 25 Dec 2016

Global CCS Institute (2016f) Uthmaniyah CO$_2$-EOR demonstration project. https://www.globalccsinstitute.com/projects/uthmaniyah-co2-eor-demonstration-project. Accessed 25 Dec 2016

Global CCS Institute (2016g) Abu Dhabi CCS project (phase 1 being Emirates Steel Industries (ESI) CCS project). https://www.globalccsinstitute.com/projects/abu-dhabi-ccs-project-phase-1-being-emirates-steel-industries-esi-ccs-project. Accessed 25 Dec 2016

Global CCS Institute (2016h) Petrobras Santos Basin pre-Salt oil Field CCS project. https://www.globalccsinstitute.com/projects/petrobras-santos-basin-pre-salt-oil-field-ccs-project. Accessed 25 Dec 2016

Gorecki CD et al (2014) Modeling and monitoring associated CO_2 storage at the Bell Creek field. Joint IEAGHG modelling and monitoring network meeting. Available via http://ieaghg.org/docs/General_Docs/1_Comb_Mod_Mon/Gorecki_13SEC.pdf. Accessed 24 Dec 2016

Grigg RB et al (2012) Phase II final scientific/technical report. Southwest Regional Partnership on Carbon Sequestration. New Mexico Institute of Mining and Technology, Socorro, New Mexico. Available via https://www.southwestcarbonpartnership.org/download/the-swp/phase-ii/PhaseII_Final_SWP_2.pdf. Accessed 24 Dec 2016

Han WS (2010) Integrated modeling approaches: understanding geologic CO_2 sequestration processes. Southwest Partnership CO_2 Sequestration, the University of Utah. Available via http://training.southwestcarbonpartnership.org/links/integ_model_approach%20.pdf. Accessed 24 Dec 2016

IEA (2015b) Storing CO_2 through enhanced oil recovery (combining EOR with CO_2 storage (EOR +) for profit. International Energy Agency. Available via https://www.iea.org/publications/insights/insightpublications/Storing_CO2_through_Enhanced_Oil_Recovery.pdf. Accessed 22 Jan 2017

Li W, Schechter DS (2014) Using polymers to improve CO_2 flooding in the North Burbank Unit. Canadian Energy Tech Inno 2(1):1–8

Meng J (2015) Fracture architecture of the high plains aquifer, northeastern Texas panhandle: implications for geologic storage of carbon dioxide. MS Thesis. Oklahoma State University. Available via https://shareok.org/bitstream/handle/11244/45366/Meng_okstate_0664M_14273.pdf?sequence=1. Accessed 25 Dec 2016

Miller J et al (2014) Alternative conceptual geologic models for CO_2 injection in a Niagaran pinnacle reef oil field, northern Michigan, USA. Energy Procedia 63:3685–3701

MIT CC&ST (2016) Carbon capture & sequestration project database. Carbon Capture and Sequestration Technologies at MIT. http://sequestration.mit.edu/tools/projects/index.html. Accessed 24 Dec 2016

NETL (2015c) Carbon storage atlas fifth edition. The U.S. Department of Energy's (DOE) National energy technology laboratory. Available via https://www.netl.doe.gov/File%20Library/Research/Coal/carbon-storage/atlasv/ATLAS-V-2015.pdf. Accessed 24 Dec 2016

Pan F et al (2016) Uncertainty analysis of carbon sequestration in an active CO_2-EOR field. Int J of Greenhouse Gas Con. 51:18–28

Riding JB, Rochelle CA (2005) The IEA Weyburn CO_2 monitoring and storage project. British Geological Survey. Research report RR/05/03 54 pp. Available via http://nora.nerc.ac.uk/3682/1/RR05003.pdf. Accessed 24 Dec 2016

White MD et al (2014) Numerical simulation of carbon dioxide injection in the western section of the Farnsworth Unit. Energy Procedia 63:7891–7912

Whittaker SG (2005) Geological characterization of the Weyburn field for geological storage of CO_2: summary of Phase I results of the IEA GHG Weyburn CO_2 monitoring and storage project. In: summary of investigations. Saskatchewan Geological Survey. Available via http://publications.gov.sk.ca/documents/310/88854-Whittaker_2005vol1.pdf. Accessed 25 Dec 2016

Whittaker SG et al (2011) A decade of CO_2 injection into depleting oil fields: monitoring and research activities of the IEA GHG Weyburn-Midale CO_2 monitoring and storage project. Energy Procedia 4:6069–6076

Wood JR et al (2006) Implementing a novel cyclic CO_2 flood in Paleozoic reefs. Michigan Technological University. Available via https://www.osti.gov/scitech/servlets/purl/888554-vPjhsg/. Accessed 24 Dec 2016

Chapter 3
Monitoring of Injected CO$_2$

Abstract It is the monitoring of injected CO$_2$ that makes a commercial CO$_2$-EOR project a storage project as well. Over the years, operators and researchers have developed, tested, and delayed various monitoring, verification, and accounting (MVA)/monitoring, management, and verification (MMV) techniques, methodologies, and practices for pre-injection, during injection, and post-injection monitoring of CO$_2$ at a storage site. Apart from a brief discussion on currently used MVA/MMV technologies, their benefits and associated challenges, best practices, emerging MVA/MMV technologies and integration of collected monitoring data and analysis are also discussed.

3.1 Introduction

The petroleum industry has long been used various surveillance and monitoring techniques like well logging and downhole monitoring, seismic and other geophysical methods, fluid sampling and tracer analysis, and even intelligent monitoring networks for effective reservoir management of commercial CO$_2$-EOR projects. However, it is the inclusion of additional pre-injection (baseline), during injection, and post-injection (site reclamation) activities executed in the form a dedicated monitoring, verification, and accounting (MVA) or measurement, monitoring, and verification (MMV) program that transforms a commercial CO$_2$-EOR project into a simultaneous CO$_2$-EOR and storage project.

In case of a reservoir surveillance and monitoring program for commercial CO$_2$-EOR project, engineers are more interested in identifying and tracking the transition zone(s) between CO$_2$ and oil, miscible front, and unswept area(s). However, MVA/MMV program for a simultaneous CO$_2$-EOR and storage project will have a broader scope and greater flexibility to ensure that the 99 percent of injected CO$_2$ remains in the designated trap [i.e., no spill out of the reservoir (injection zone(s)] while providing the typical EOR related monitoring support to the engineers.

To ensure that injected CO$_2$ remains in the targeted injection zone(s) and that injection wells and preexisting wells are not prone to unintended CO$_2$ release,

© The Author(s) 2017
D. Saini, *Engineering Aspects of Geologic CO$_2$ Storage*, SpringerBriefs
in Petroleum Geoscience & Engineering, DOI 10.1007/978-3-319-56074-8_3

surface, near-surface, and subsurface monitoring techniques and methods can be deployed and integrated into MVA data integration and analysis technologies (NETL 2012). There are all sorts of MVA/MMV technologies out there, some are commercial and some are either in the early demonstration stage or still in the development stage, however, some of the monitoring technologies are best-suited for meeting the regulatory compliance needs (primary technologies) and some of them are better suited for effective management of the reservoir (secondary technologies). Kuuskraa et al. (2013) recently provided a detailed discussion on the synergistic pursuit of advances in MMV technologies for simultaneous CO$_2$-EOR and storage projects.

Both primary and secondary monitoring technologies are the technologies that have been extensively used by the petroleum industry for reservoir surveil-lance and monitoring in the EOR projects. The primary technologies are considered proven in demonstrating 99% containment of CO$_2$ (i.e., injected CO$_2$ remains entirely in the subsurface over significant time spans (hundreds to thousands of years) (NETL 2009). Secondary technologies often help in site characterization to support site-specific geologic and reservoir simulation modeling studies and require additional demonstration that they are sufficiently precise and quantitative to detect, locate, and quantify emissions from a storage site (NETL 2009).

3.2   Currently Used MVA/MMV Technologies

Table 3.1 provides a snapshot of currently used MVA/MMV technologies at operational LSIPs and other large-scale simultaneous CO$_2$-EOR and storage projects. Commercially available subsurface monitoring technologies such as time-lapsed surface 3-D seismic (seismic (four-dimensional (4-D) seismic)), time-lapsed 3-D vertical seismic profile (four-dimensional VSP (4-D VSP)), pressure, temperature, and flow rate monitoring, and geochemical/tracer tests, pulse neutron capturing (PNC) logging are the most commonly used MVA/MMV tools in these projects. Wellhead pressure monitoring is another subsurface monitoring technique that have been used by these projects. The above zone-pressure monitoring has also been used as a subsurface monitoring tool at both the Bell Creek and West Hastings storage sites. The microgravity, microseismic and borehole gravity are the other subsurface monitoring technologies that have been used in these projects.

The shallow groundwater sampling is most used near-surface monitoring technique that has been employed by these projects. The soil-gas sampling appears to be the monitoring technique of choice for near-surface and surface monitoring, should injected CO$_2$ migrate out of the designated trap. The Interferometric Synthetic Aperture Radar (InSAR), another near-surface monitoring technique, has also been used at one project.

Table 3.1 Key monitoring techniques employed at storage sites of some of the currently operational North American LSIPs and other large-scale simultaneous CO$_2$-EOR and storage projects

Technique	Monitoring category	Storage site*						
		Weyburn Oil Unit	Bell Creek	SACROC Unit	West Hastings	Pinnacle Reefs (Michigan's Northern Reef Trend)	Farnsworth unit
Atmospheric CO$_2$ and/or methane fluxes	Atmospheric/surface	–	–	–	–	–	X
Soil-gas sampling	Surface/near-surface	–	X	X	X	–	X
Surface water sampling	Surface	–	X	–	–	–	X
Self-potential	Surface	–	–	–	–	–	X
Passive seismic	Surface	X	X	–	–	–	X
Shallow groundwater sampling	Near-surface	X	X	X	X	–	X
Interferometric Synthetic Aperture Radar (InSAR)	Near-surface	–	–	–	–	X	–
Pressure, temperature, and flow rate monitoring	Subsurface	X	X	–	X	X	X
Wellhead pressure	Subsurface	–	X	–	X	–	X
Pulse Neutron Capture Logging (PNC/PNL)	Subsurface	–	X	X	–	X	–
Pulse Neutron Spectroscopy (PNS) Logging	Subsurface	–	–	X	–	–	–
Microseismic	Subsurface	–	–	–	–	X	X
Time-lapsed surface 3-D seismic (4-D seismic)	Subsurface	X	X	X	X	–	X

(continued)

Table 3.1 (continued)

Technique	Monitoring category	Storage site*						
		Weyburn Oil Unit	Bell Creek	SACROC Unit	West Hastings	Pinnacle Reefs (Michigan's Northern Reef Trend)	Farnsworth unit
Time-lapsed 3-D Vertical Seismic Profiling (VSP) (4-D VSP)	Subsurface	–	X	X	X	X	X
Geochemical/Tracer Tests	Subsurface	X	–	X	X	X	X
Borehole gravity survey	Subsurface	–	–	–	–	X	–
Microgravity survey	Subsurface	–	–	–	X	–	X
Above zone-pressure monitoring	Subsurface	–	X	–	X	–	–

*Data sources include Balch et al. (2016), Gerst et al. (2014), Gorecki (2015), Grigg et al. (2012), Mehlman (2010), NETL (2012, 2015c), Whittaker et al. (2011)

The passive seismic and surface water sampling are the only surface monitoring techniques that have been used these projects. The Farnsworth Unit project is the only project where atmospheric monitoring of CO_2 fluxes has also been included in the MVA/MMV program.

3.3 Benefits and Challenges of Currently Used MVA/MMV Technologies

There is no doubt that periodic surface 3-D seismic survey, a mature technology, has been used by majority of the projects (listed in Table 3.1) for obtaining high-quality information on spatial distribution and migration of CO_2. However, it is of semi-quantitative nature and cannot be used for mass-balance CO_2 dissolved or trapped as mineral (NETL 2009).

Another mature technology, namely, time-lapsed 3-D VSP, provides robust information on CO_2 concentration and migration and because it involves the use of a single wellbore, its resolution is more than surface seismic (NETL 2009). Its use for calibrating the 3-D surface seismic survey makes it a compliment technique. That is why, both techniques have been included in MVA/MMV programs of majority of the projects (Table 3.1). However, wellbore geometry (vertical well vs. directional well) can limit its application (NETL 2009).

The pulse neutron capture (PNC) logging is a wireline monitoring technology that uses pulsed neutron techniques for depicting phase saturations (oil, gas, water), lithology, and porosity (NETL 2009). It is a high-resolution tool that can identify specific geology parameter around the wellbore, however, limited radius of investigation around the wellbore is the biggest challenge while using this technique. In time-lapse mode, it provides a quantitative information on CO_2 saturation, however, wellbore conditions such as fluid invasion due to workover can affect its accuracy and the presence of low salinity formation water can make it less sensitive (NETL 2009).

All three techniques discussed above are the secondary monitoring technologies. It means, they are routine technologies that have been extensively used by the petroleum industry for reservoir surveillance and monitoring in the EOR projects. However, repeated use of these technologies (time-lapse) can be quite expensive.

Groundwater monitoring (near-surface) and wellhead pressure monitoring (subsurface) are the two commonly used primary monitoring technologies that have been used by majority of the projects listed in Table 3.1. Sampling of water for basic chemical analysis is a mature technology that provides easier detection that atmospheric monitoring tools and early detection prior to large emissions (NETL 2009). However, for null results (no CO_2 leakage), repeated and frequent sampling efforts can be significant and proved costly in long run.

Well head (annulus) pressure monitoring, a reliable testing method with simple equipment, can be done constantly to test the mechanical integrity of the wellbore for detecting leakage from the casing, packer, or tubing (NETL 2009). However, periodic testing requires stopping the injection process during testing thus poses an additional operational challenge. Soil and Vadose Zone gas monitoring, one of the most often used near-surface/surface monitoring technique in the projects listed in Table 3.1, is a secondary technology that relies of sampling of gas in vadose zone/soil (near surface) for any CO_2 leakage. Detection of elevated CO_2 concentrations well above background levels (baseline surveys) provides indication of leak and migration from the target reservoir, however, significant effort for null result (no CO_2 leakage) and relatively late detection of leakage pose challenge in the use of this technology (NETL 2009).

Eddy Covariance is the only atmospheric monitoring technique that has been used in any of the projects (Farnsworth Unit project (Balch et al. 2016)) for measuring atmospheric CO_2 concentrations at a height above the ground surface. It is a mature secondary technology that can provide accurate data under continuous operation, however, it requires very specialized equipment and robust data processing (NETL 2009).

3.4   Best Practices

The Best Practice Manual (BPM) prepared by the NETL (2012), provide a detailed discussion on the diverse portfolio of effective MVA tools and technologies that is needed to meet the needs of individual geologic storage project and to support its widespread commercial deployment by 2030. The BPM (NETL (2012) also provides an over of existing MVA technologies; a discussion on the maturity of field readiness of these technologies; a discussion on the suitability of various monitoring tools to meet specific project needs for regulatory compliance and/or reservoir management; and guidelines for developing and executing an MVA plan/program.

A robust MVA/MMV program will definingly include both aspects (regulatory compliance and reservoir management) of the monitoring of injected CO_2 in a simultaneous CO_2-EOR and storage project. Clearly the ideal monitoring technology would be the one that can directly measure the mass of CO_2 stored via measurement of the spatial extent of CO_2 in the subsurface, CO_2 saturation and CO_2 density (Boait et al. 2015). In the absence of direct measurement of stored mass of CO_2 in the target trap(s) [injected zone(s)], an MMV program would aim at identifying migration within the storage complex and leakage out of the complex (Boait et al. 2015). However, there is hardly a single monitoring technique that could achieve the stated objectives of a MVA/MMV program.

3.5 Emerging MVA/MMV Technologies

Monitoring of injected CO_2 over an extend time spans may be a costly affair. As stressed by Boait et al. (2015), the largest differences in monitoring technology usage is not process related, rather site specific geology and geography control it. Per them (Boait et al. 2015) the monitoring technology uses are shown to be largely related to the level of characterization, baseline assessment, likely infrastructure in place and pressure management during operations. Further, pressure and volume reduction/increase may complicate monitoring of injected CO_2 due to reasons like formation of mineral phases or lose of reservoir and/or mechanical strengths of reservoir (injection zone) and overlying seals (cap rocks) or cycles of depletion and re-pressurizations which is a common occurrence in depleted oil fields.

As documented by Li et al. (2016) and Litynski et al. (2013), currently used monitoring technologies have their own application limitations. Several remote sensing technologies, like InSAR have high potential to supplement other technologies for cost-effective, repeatable, reliable, and efficient long-term monitoring of injected CO_2 because they can perform under non or minimally intrusive conditions; can efficiently obtain data from relatively large areas with low relative effort; can be used as a cost-effective alternative to other options; and can be used in special applications where other forms of MVA may not be feasible (Litynski et al. 2013).

The development of technologies such as an automated high power permanent low-cost automated borehole seismic source system comprising of two downhole seismic sources: one for high-resolution cross-well surveys and one for high-resolution vertical seismic profile surveys, or a low-impact Krauklis wave (K-wave) monitoring method that may potentially eliminate several shortcomings of traditional seismic monitoring methods, such as high cost, disruptive surface impacts, and long intervals between surveys, while providing timely, actionable information to the field operator appear to potential and emerging MVA technologies (U.S. DOE 2016).

Recently developed pocket-sized gravimetry technique (AAPG 2016), if could be integrated into existing 4-D surface microgravity survey techniques while enabling high precision measurements, also appears to hold a potential for becoming a low-cost MVA technology in near future.

3.6 MVA/MMV Data Integration and Analysis

Finally, MVA/MMV data integration and analysis technologies such as intelligent monitoring networks and advanced data integration and analysis software (NETL 2012) are needed for better understanding, integration, and analysis of wide variety of monitoring data.

An intelligent monitoring network may combine data from CO_2 monitoring wells, surface monitoring sensors, subsurface monitoring tools, and injection equipment and the data are compiled in real time in a database that is updated continuously (NETL 2012). One of the examples intelligent monitoring network is the application of SmartWell technology (i.e., installation of downhole flow-control volumes and isolation packers) in a five-well research project associated with the CO_2 flood in the SACROC Unit (Kelly Snyder oil field), West Texas (Kuuskraa et al. 2013). By implementing SmartWell Technology at SACROC, operator was able to shut-off CO_2 breakthrough in a producing well without the expense or formation damage of a workover (Kuuskraa et al. 2013).

In addition to intelligent monitoring networks (IMS), scientists are developing advanced software tools such as data processing software data as well as integration and visualization software to perform specific MVA data integration and analysis tasks (NETL 2012). However, creation of advanced, integrated measurement and control systems to track CO_2 before, during, and after injection and improve injection efficiency and development of new, real-time, data-capable workflows, algorithms, and user interfaces that will automate the integration of monitoring data and simulations as part of an IMS at CO_2 storage sites, which will allow more efficient operations in the context of a site's evolving risk profile, optimize storage efficiency and capacity, and guide cost-effective MVA efforts (NETL 2016) will going to take some time before becoming commercially available.

References

AAPG (2016) New tech: pocket-sized gravimetry. American Association of Petroleum Geologists. Available via http://www.aapg.org/publications/news/explorer/emphasis/Articleid/34012/new-tech-pocket-sized-gravimetry. Accessed 25 Jan 2016

Balch R et al (2016) Monitoring CO_2 at an enhanced oil recovery and carbon capture and storage project, Farnsworth Unit, Texas. In: Krutka H et al (eds) ECI symposium series. Available via http://dc.engconfintl.org/co2_summit2/30. Accessed 25 Jan 2016

Boait F et al (2015) Measurement, monitoring, and verification: enhanced oil recovery and carbon dioxide storage. Scottish Carbon Capture & Storage. Available via http://www.sccs.org.uk/images/expertise/misc/SCCS-CO2-EOR-JIP-MMV.pdf. Accessed 26 June 2016

Gerst J et al (2014) Using baseline monitoring data to strengthen the geological characterization of a Niagaran Pinnacle Reef. Energy Procedia 63:3923–3934

Gorecki CD (2015) The Plains CO_2 Reduction (PCOR) Partnership: Bell Creek field project. Carbon storage R&D project review meeting. Available via https://www.netl.doe.gov/File%20Library/Events/2015/carbon%20storage/proceedings/08-19_05_Gorecki.pdf. Accessed 25 Jan 2017

Grigg RB et al (2012) Phase II final scientific/technical report. Southwest regional partnership on carbon sequestration. New Mexico Institute of Mining and Technology, Socorro, New Mexico. Available via https://www.southwestcarbonpartnership.org/download/the-swp/phase-ii/PhaseII_Final_SWP_2.pdf. Accessed 24 Dec 2016

Kuuskraa VA et al (2013) The synergistic pursuit of advances in MMV technologies for CO_2—enhanced recovery and CO_2 storage. Energy Procedia 37:4099–4105

Li Q et al (2016) Monitoring of carbon dioxide geological utilization and storage in China: a review. In: Wu Y et al (eds) Acid gas extraction for disposal and related topics. Wiley-Scrivener, New York, pp 331–358

Litynski J et al (2013) U.S. Department of Energy efforts to advance remote sensing technologies for monitoring geologic storage operations. Energy Procedia 37:4114–4127

Mehlman S (2010) Carbon capture and sequestration (via enhanced oil recovery) from a hydrogen production facility in an oil refinery. U.S. Department of Energy, National Energy Technology Laboratory. Available via http://www.osti.gov/scitech/servlets/purl/1014021. Accessed 25 Jun 2016

NETL (2009) Monitoring, verification, and accounting of CO_2 stored in deep geologic formation. The U.S. Department of Energy's (DOE) National Energy Technology Laboratory. First addition, DOE/NETL-311/081508. Available via http://www.egcfe.ewg.apec.org/projects/EWG052010A/References/NETL_Best%20Practices_MVA.pdf. Accessed 26 Jan 2017

NETL (2012) Best practices for monitoring, verification, and accounting of CO_2 stored in deep geologic formation. The U.S. Department of Energy's (DOE) National Energy Technology Laboratory. Second addition, DOE/NETL-2012/1568. Available via https://www.netl.doe.gov/File%20Library/Research/Carbon%20Seq/Reference%20Shelf/MVA_Document.pdf. Accessed 8 Jan 2017

NETL (2015c) Carbon storage Atlas fifth edition. The U.S. Department of Energy's (DOE) National Energy Technology Laboratory. Available via https://www.netl.doe.gov/File%20Library/Research/Coal/carbon-storage/atlasv/ATLAS-V-2015.pdf. Accessed 24 Dec 2016

NETL (2016) Intelligent monitoring. The U.S. Department of Energy's (DOE) National Energy Technology Laboratory. https://www.netl.doe.gov/research/coal/carbon-storage/research-and-development/intelligent-monitoring. Accessed 26 Jan 2017

U.S. DOE (2016) DOE investing $11.5 million to advance geologic carbon storage and geothermal exploration. U.S. Department of Energy (Office of Fossil Energy). https://energy.gov/under-secretary-science-and-energy/articles/doe-investing-115-million-advance-geologic-carbon. Accessed 27 Jan 2017

Whittaker SG et al (2011) A decade of CO_2 injection into depleting oil fields: monitoring and research activities of the IEA GHG Weyburn-Midale CO_2 monitoring and storage project. Energy Procedia 4:6069–6076

Chapter 4
Injection Strategies

Abstract The currently operational simultaneous CO_2-EOR and storage projects have mainly relied on petroleum industry's experience with conventional water alternating gas (WAG) injection and continuous CO_2 injection strategies for recovering additional oil while storing the injected CO_2 in the reservoir. Hence, an improved understanding of key reservoir engineering considerations that have significant influence on the success of these currently used injection strategies can help the industry professionals in optimizing both aspects (EOR and storage) of the project. The same can also be used for evaluating the potential of alternative injection strategies such as gas assisted gravity drainage (GAGD) and formation water extraction.

4.1 Currently Used Injection Strategies

The conventional WAG injection and continuous CO_2 injection, both in miscible mode, are the two injection strategies that operators have mainly utilized in currently operational LSIPs and other simultaneous CO2-EOR and storage projects. In conventional WAG, a predetermined volume of CO_2 is injected in cycles alternating with equal volumes of water (Verma 2015). Typical cycle times range from months to a year, and the volumetric ratio of water to dense CO_2 is commonly1:1–2:1 (Enick et al. 2012).

The petroleum industry has preferred the WAG process because it helps in reducing the mobility of CO_2 in the reservoir. The injection of water along with the CO_2 helps in overcoming the gas override and reduces the CO_2 channeling thereby improving overall CO_2 sweep efficiency (Verma 2015). Also, unlike water, CO_2 is not a cheap injection fluid, hence, use of a WAG injection strategy often helps the operators in keeping the project cost down and in increasing the project life. However, these factors only paint a partial picture of the overall oil recovery factor (RF), which is the focus of a CO_2-EOR project after primary and/or secondary waterflooding recoveries. Traditionally, depleted oil fields, after primary depletion of natural reservoir energy, undergo secondary waterflooding (pressure maintenance)

© The Author(s) 2017
D. Saini, *Engineering Aspects of Geologic CO2 Storage*, SpringerBriefs
in Petroleum Geoscience & Engineering, DOI 10.1007/978-3-319-56074-8_4

for recovering additional oil, however, a CO_2-EOR project can also be launched immediately after the primary depletion.

Although the use of WAG injection strategy yields better results that continuously injecting CO_2, WAG may still leave lot of oil (approximately one third to two-thirds of the oil left behind by waterflooding) behind (Enick et al. 2012). On the other end, a WAG injection strategy will essentially result in less CO_2 stored in the reservoir compared to a continuous CO_2 injection in miscible mode. From storage point of view, use of WAG results in the reduction of pore space that may be otherwise available for injection CO_2. Also, the projects not optimized for storage, may only store up to 50–60% of the total emissions resulting from the combustion of oil/gas produced in the project and the energy consumption and other operation emissions occurring during in the entire process of producing them (reservoir to the end use point). However, optimization of injection strategies can assist us in making the EOR and storage projects "net zero emissions" projects (i.e., storing more CO_2 than the CO_2 generated by the energy consumption, operational emissions, and the end use (combustion) of the produced oil/gas).

Let's revisit some of the basic reservoir engineering considerations behind currently used injection strategies that play a critical role in determining the oil recovery (EOR) and, thus, storage capacity (storage) of a simultaneous CO_2-EOR and storage project.

4.1.1 Overall Oil Recovery Factor

Basically, for a simultaneous CO_2-EOR and storage project, overall recovery factor (RF), defined as the volume of oil recovered over the volume of original oil in place (OOIP), can be described as a product of following four efficiency terms (Muggeridge et al. 2014):

1. The macroscopic sweep efficiency, E_S, which is the fraction of the connected reservoir volume that is swept by the injected fluids.
2. The microscopic displacement efficiency, E_{PS}, described as the fraction of oil displaced from the pores by the injected fluids (water and/or CO_2), in those pores which are contacted by the injected fluids.
3. The connected volume factor, E_D, represents the proportion of the total reservoir volume connected to the wells.
4. The economic efficiency factor, E_C, which imposes additional physical and commercial constraints on the project life.

Hence, an equation for overall RF can be written as:

$$RF = E_S \times E_{PS} \times E_D \times E_C \qquad (3.1)$$

Because, each of the terms (efficiency or factor) in the Eq. 3.1 represent a fractional quantity (<1), hence, to maximize the RF, these four terms need to be increased near to 1.

4.1.1.1 Macroscopic Efficiency, E_S

The macroscopic sweep efficiency, E_S, of the reservoir is difficult to quantify if the reservoir does not have a previous history of waterflooding or gas (CO_2, hydrocarbon, N_2 etc.) injection. Nevertheless, E_S, is principally affected by the geological heterogeneity in the reservoir. One common and adverse manifestation of geological heterogeneity is the presence of high permeability channels or layers (Muggeridge et al. 2014) in the injection zone. These high permeability layers may be the result of geological deposition history of the formation, or a product of fractures, natural or man-made (Meyer 2007). The injected water will preferentially flow through the high permeability layers, thus, bypassing volumes of oil present in low permeability layers. On the other end, gravitational segregation, arising due to the higher density contrast between gas and oil compared to oil and water, will cause injected CO_2 to rise above the oil and then flow rapidly along the top of the reservoir.

Another key factor that exacerbates the effect of geological heterogeneity is the mobility of injected fluids (water and/or CO_2) in the presence of displaced fluid (oil). The Mobility Ratio, M, is often described as the mobility of the displaced fluid and the mobility of displacing fluid in the porous medium. A mathematical expression for M is given by:

$$M = \frac{\left(\dfrac{k_r}{\mu}\right)_{displacing fluid}}{\left(\dfrac{k_r}{\mu}\right)_{displaced fluid}} \tag{3.2}$$

Where k_r is the relative permeability term, and μ is the viscosity of a fluid phase [displacing (injected fluid) and displaced (reservoir oil)]. Generally, displaced fluid (oil) has higher viscosity compared to displacing (injected water), thus, resulting in an unfavorable mobility ratio (M > 1). It causes viscous figuring of injected water which leads to early water breakthrough at production well and a low E_S. If the injected fluid is CO_2, viscous figuring problem will be more serious as viscosity of displaced fluid (oil) will be still much higher despite that fact that the mixing of injected CO_2 with oil results in swelling of oil resulting in a viscosity reduction of oil phase. The occurrence of viscous fingering leads to early CO_2 breakthrough, lower oil production rates, and high CO_2 utilization ratios (i.e., need of higher injected CO_2 volumes for recovering per barrel of oil), all contributing to a low E_S.

In most cases, it is the preferential flow of injected fluid through higher permeability layers rather than viscous fingering that dominates macroscopic sweep efficiency. One common way to reduce viscous fingering problem is near-wellbore conformance control, which is often achieved by applying gel treatments to block preferential flow paths (high permeability layers, fractures). CO_2 conformance control foams are also an alternative technology (Enick et al. 2012) for conformance control in CO_2 flooding.

Viscous fingering can be suppressed or reduced by either altering the relative permeability or modifying viscosity resulting in a more favorable ratio ($M \leq 1$). The technique, commonly referred as mobility control, most readily accomplished using WAG injection strategy i.e., the injection of both CO_2 and water in an alternating sequence. The alternating injection of water (brine) and CO_2 results in an increased water saturation and reduced CO_2 saturation within the pores (Enick et al. 2012). Another option is to change the CO_2 viscosity itself. A CO_2 thickener would be a particularly effective mobility control agent due to its ability to manipulate the viscosity of the injected CO_2. As stressed by Enick et al. (2012), an affordable CO_2 thickener can be a Game-Changing technology, however, it has not yet been developed for commercial use.

Gravitational segregation is another main reason responsible for a low E_S that arises from the density contrast between the injected and the displaced fluids. Obviously, density contrast is more profound in CO_2-oil rather than water-oil displacements. Due to its relatively low density and viscosity compared to the densities and viscosities of reservoir fluids (oil and water), injected CO_2 tends to rise and then flow rapidly along the top of the reservoir in an unstable gravity tongue (Fayers and Muggeridge 1990 in Muggeridge et al. 2014). The use of top-down CO_2 injection in miscible mode at the Weyburn-Midale LSIP is an excellent example of the reservoir engineering solutions devised by petroleum industry for minimizing the impact of gravity segregation on E_S.

4.1.1.2 Microscopic Efficiency, E_{PS}

The typical microscopic sweep efficiency, E_{PS}, from a water flood is 70% or less (Muggeridge et al. 2014). Both the pore-scale capillary effects and the relative permeability characteristics of the rock affect E_{PS}. A measure to quantify the influence of pore-scale capillary effect is capillary number, Ca, which is manifestation of delicate interplay of viscous, capillary, and, rock/oil adhesion forces. A generalized equation for capillary number can be written as:

$$Ca = \frac{v\mu}{\sigma \cos \theta} \qquad (3.3)$$

Where v is the interstitial velocity of fluid in porous medium (derived from Darcy's Law), μ is the fluid viscosity, σ is the interfacial tension (IFT) between the displaced and displacing fluid, and θ is the contact angle, a measure of wettability (in other words, the extent of rock/oil adhesion force present in the system) of the reservoir.

The phenomenon of capillary trapping (i.e., trapping of a non-wetting phase in porous medium as discontinuous, non-moveable pore-scale clusters due to capillary forces) greatly influences the ability of flow of various one fluid phase in the presence of another fluid phase through interconnected small-size pores of reservoir when Ca is smaller than 10^{-5}. It is very difficult to significantly increase the interstitial velocity of fluids even a sufficiently large pressure gradient is available between the injection and production wells. Also, it will be difficult to maintain this velocity due while injecting a high-viscosity fluid (e.g., CO_2 thickener). The size (pore diameter) of interconnected pores in porous medium, itself, greatly magnifies the magnitude of viscous force, thus, snapping the continuous phase and creating discontinuous ganglia that is trapped in the pore space.

Because, capillary trapping is primarily controlled by the flow, rock/fluids interactions, and the pore structures, only practical way to increase the capillary number and, thus, reducing capillary trapping effect and pushing the trapped ganglia towards the production well, is to minimize the magnitude of rock/fluids interactions (IFT and wettability). Further, it is more convenient to reduce the IFT between the oil and the injected fluid via attainment of miscibility between oil and CO_2.

Often, $\cos \theta$ term is assumed to be 1 (i.e., $\theta = 0°$) while evaluating the effect of capillary number on E_{PS}. This approximation generally holds for water-wet system, where no or negligible rock/oil adhesion force is present. However, majority of currently producing miscible CO_2-EOR projects in the US are in carbonate reservoirs and carbonate reservoirs are known for their oil-wet (i.e., presence of strong rock/oil adhesion force) tendencies. Readers are referred to the published articles such as Rao (2003) and Rao and Maini (1993) for detailed discussions on the role and impact of rock/oil adhesion interactions on reservoir mechanics.

In case of oil-wet reservoirs, the occurrence of discontinuous oil ganglia due to capillary effect is significant reduced as oil stays as a continuous bulk phase next to the rock surface compared to water-wet reservoirs where water stays as a continuous bulk phase next to the rock surface. However, injected fluid, now, just flows through the middle of the pores and the pore throats without significantly contacting the oil staying next to the rock surface and resulting in an early breakthrough. As increasingly fluid is injected, it is possible to get more contact between the oil and injected fluid, however, it will take a long period and a very large throughput of injected fluid, to push majority of the oil (staying next to the rock surface) toward the production well.

4.1.1.3 Contact Volume Factor, E_D, and Economic Efficiency Factor, E_C

As explained by Muggeridge et al. (2014), E_D, represents the proportion of the total reservoir volume in communication through production and injection wells. Certain geological features like faults and low permeability barriers (shale streaks) may results in isolation of a portion of reservoir volume that cannot be accessed via exiting wells. Hence, first, the identification of non-swept portions of the reservoir by techniques such as tracer and seismic surveying and then, drilling of infill wells to sweep the oil out from those portions of the reservoir is a practical way to improve E_D.

An improvement in E_C can be achieved by implementing the prudent reservoir and production management practices that will result in extending project life (i.e., achieving economically sustainable oil production rates over a longer period), lesser water and gas production volumes, extended life spans of pipelines, production systems, and other surface facilities.

4.1.2 Storage Capacity

Recovering maximum oil while keeping all the injected CO_2 in the reservoir is a challenge to any simultaneous CO_2-EOR and storage project. An operator always aims for the use of minimum mass (or volume) of CO_2 to recover a barrel of oil. However, for a simultaneous CO_2-EOR and storage project, it is equally important that the project can inject a desired amount of CO_2 and produces a minimal portion of the injected CO_2 back to the surface. Typically, any of the produced CO_2 is processed (i.e., separated from production fluids (oil and water), dried, re-compressed, and re-injected back either in the same reservoir (another injection area/phase) or in another project nearby. There are certain approaches have been suggested in the published literature (Jessen et al. 2005; Advanced Resources International (ARI) and Melzer Consulting 2010; Hill et al. 2013; IEA 2015b) for increasing CO_2 storage in oil recovery.

One of the options is to optimize the water injection (timing, injection rates, and WAG ratio) to minimize gas cycling and maximum gas storage. Another option is not only considering reservoir re-pressurization after the end of the producing life of the field, but also, the candidate depleted reservoir can also be re-pressurized via CO_2 injection instead of raising reservoir pressure above MMP via waterflooding. Reservoir re-pressurization during the disposal of acid gas (67% CO_2 + 33% H_2S) resulting from natural gas processing, had resulted in recovering additional oil from Zama F Pool (Saini et al. 2013), and led the operator to launch a formal project to for demonstrating the viability of simultaneous CO_2-EOR and storage project in closed pinnacle reef structure.

In a scenario of the availability of low cost CO_2, re-pressurization via CO_2 injection can be an excellent injection strategy for both enhanced oil recovery and enhanced CO_2 storage purposes. Denbury Resources' experience with deploying CO_2 injection earlier in field development is an excellent example of eliminating the need of water injection during injection phase and substantially increasing the amount of CO_2 stored in the reservoir. The company has implemented its continuous CO_2 injection (miscible mode) strategy in the Bell Creek oil field

(Lost Cabin LSIP), however, reservoir re-pressurization was achieved by waterflooding. In Denbury's continuous miscible injection strategy, production wells are allowed to flow on their own by maintaining the reservoir pressure higher than the normal hydrostatic pressure of the reservoir. Also, any of the re-cycled CO_2 is repeatedly injected back into the reservoir. The self-flowing production wells results in significant oil lifting cost savings that offset the additional costs to compress and re-cycle CO_2 through the reservoir (Merchant 2015). However, such strategy appears to works well in sandstone reservoirs that have a high injection flow capacity.

Another strategy will be not re-injecting produced water and recycled CO_2 into the reservoir back but injecting any of the recycled CO_2 into a saline aquifer while disposing the produced water into an exempted (non-USDW) aquifer. Injection of CO_2 for recovering oil from the "residual oil zones" (ROZs) that may be encountered in naturally water-flooded intervals below the established oil-water contacts (Hill et al. 2013) also appears to be a potential way for enhancing CO_2 storage in EOR projects.

Reduction in the adverse effects of preferential flow of injected CO_2 through high permeability layers via partial completion of both injection and production wells as well as the use of horizontal wells to distribute the injected CO_2 in larger reservoir volume and using it to produce oil from this extended reservoir volume. For improving the contact between injected CO_2 and reservoir oil, injections wells can be partially completed in bottom part of the injection zone rather than the entire zone. Similarly, a production well completed in bottom part of the injection zone will also delay CO_2 breakthrough and reduce its production as injected CO_2, due to gravity segregation, will rise above the oil column and in the prosses improve the gravity assisted oil drainage towards the production well. Also, not completing the production wells in high permeability layer is a good completion strategy that will avoid the channeling of injected CO_2 through these layers thus increasing the retention of the injected CO_2 in the reservoir.

4.2 Alternative Injection Strategies

4.2.1 Gas Assisted Gravity Drainage (GAGD)

Apart from conventional WAG and continuous CO_2 injection in miscible mode that mainly rely on pattern flooding, an alternative injection strategy, namely, Gas Assisted Gravity Drainage (GAGD) process (Rao et al. 2004) appears to be a promising injection strategy from both oil recovery and storage capacity point of view. A review of the use of gravity drainage concept in the field shows that it is applicable to all reservoir types and reservoir characteristics using common injectant gases in both secondary and well as tertiary recovery modes (Kulkarni and Rao 2006). However, historically, gravity stable (drainage) injections has been applied to highly dipping and reef type reservoirs only.

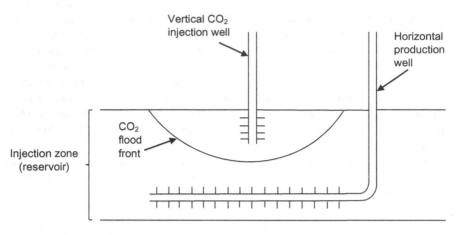

Fig. 4.1 Schematic of GAGD process (after Rao et al. 2004)

The GAGD process (Fig. 4.1) involves the placing a horizontal producer near the bottom of the productive zone and injecting gas through vertical injection wells completed in the top portion of the productive zone. As the injected gas rises to the top to forming a gas bank, oil and water drain down to the horizontal producer (Rao et al. 2004). The GAGD process uses the gravitational flow to improve the sweep efficiency in the reservoir and expends the applicability of the gravity stable gas injection concept to various types of reservoirs including horizontal productive zone. Also, a combination of vertical injectors and horizontal producers will significantly delay the CO_2 breakthrough at production wells which is a desired scenario from CO_2 storage point of view.

One of the currently operational large-scale simultaneous CO_2-EOR and storage project, namely, the Michigan Basin Project is an excellent example of gravity stable injection in a reef reservoir. However, GAGD injection strategy has not yet caught the full attention of the operators of simultaneous EOR and storage projects.

4.2.2 Formation Water Extraction

Extraction of the water from the water leg (below the established oil-water contact), if present, for creating additional pore space for injected CO_2 in a simultaneous EOR and storage project also appears to be a good way for enhancing both oil recovery and storage capacity. The numerical modeling studies such as Saini et al. (2013) conducted for evaluating the viability of top down miscible CO_2 injection coupled with formation water extraction (Fig. 4.2) have shown some promise for such alternative injection strategy. However, this strategy needs to be further explored and tested in the field for establishing it as a viable injection strategy.

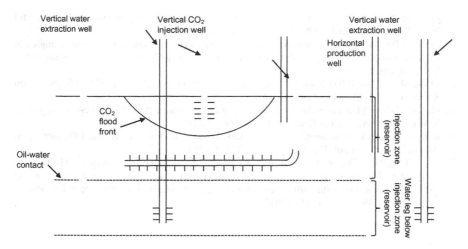

Fig. 4.2 Schematic of top down continuous CO_2 injection coupled with formation water extraction

References

ARI, Melzer Consulting (2010) Optimization of CO_2 storage in CO_2 enhanced oil recovery projects. U.S. Department of energy & climate change (DECC), Office of carbon capture & storage. Available via https://www.gov.uk/government/uploads/system/uploads/attachment_data/file/47992/1006-optimization-of-co2-storage-in-co2-enhanced-oil-re.pdf. Accessed 1 Jan 2017

Enick RM et al (2012) Mobility and conformance control for CO_2 EOR via thickeners, foams, and gels—a literature review of 40 years of research and pilot tests. Soc Pet Eng. doi:10.2118/154122-MS

Fayers FJ, Muggeridge AH (1990) Extensions to Dietz theory and behaviour of gravity tongues in slightly tilted reservoirs. SPE Reservoir Eng 5:487–494. doi:10.2118/18438-PA

Hill B, Hovorka S, Melzer S (2013) Geologic carbon storage through enhanced oil recovery. Energ Procedia 37:6808–6830

IEA (2015b) Storing CO_2 through enhanced oil recovery (combining EOR with CO_2 storage (EOR+) for profit. International energy agency. Available via https://www.iea.org/publications/insights/insightpublications/Storing_CO2_through_Enhanced_Oil_Recovery.pdf. Accessed 22 Jan 2017

Jessen K, Kovscek AR, Orr FM Jr (2005) Increasing CO_2 storage in oil recovery. Energy Convers Manag 46(2):293–311

Kulkarni MM, Rao DN (2006) Characterization of operative mechanisms in gravity drainage field projects through dimensional analysis. Society of Petroleum Engineers. doi:10.2118/103230-MS

Merchant D (2015) Life beyond 80—a look at conventional WAG recovery beyond 80% HCPV injection in CO_2 tertiary floods. Carbon Manag Technol Conf. doi:10.2118/440075-MS

Meyer JP (2007) Summary of carbon dioxide enhanced oil recovery (CO_2EOR) injection well technology. American petroleum institute. Available via http://www.api.org/ ∼ /media/files/ehs/climate-change/summary-carbon-dioxide-enhanced-oil-recovery-well-tech.pdf. Accessed 22 Jan 2017

Muggeridge A et al (2014) Recovery rates, enhanced oil recovery and technological limits. Phil
    Trans R Soc A 372:20120320. doi:10.1098/rsta.2012.0320
Rao DN (2003) The concept, characterization, concerns and consequences of contact angles in
    solid-liquid-liquid systems. In: Mittal KL (ed) Contact Angle, wettability and adhesion, vol 3.
    VSP, Utrecht, pp 191–210
Rao DN, Maini BB (1993) Impact of adhesion on reservoir mechanics. CIM-93-65. Petroleum
    Society of CIM, Calgary, Alberta
Rao DN et al (2004) Development of gas assisted gravity drainage (GAGD) process for improved
    light oil recovery. Soc Pet Eng. doi:10.2118/89357-MS
Saini D et al (2013) A simulation study of simultaneous acid gas EOR and CO_2 storage at
    Apache's Zama F Pool. Energy Procedia 37:3891–3900
Verma MK (2015) Fundamentals of carbon dioxide-enhanced oil recovery (CO2-EOR)—a
    supporting document of the assessment methodology for hydrocarbon recovery using
    CO2-EOR associated with carbon sequestration. U.S. Geological Survey. Open-File Report
    2015–1071, p 19. doi:10.3133/ofr20151071

Chapter 5
Evaluation of CO_2-Oil Miscibility

Abstract Injection of CO_2 in miscible mode is preferred in simultaneous CO_2-EOR and storage projects, hence a better understanding of CO_2-oil miscibility, mechanisms responsible for attaining CO_2-oil miscibility in the reservoir, and fast as well as robust techniques for determining the minimum miscibility pressure (MMP) is essential for successful designing of future geologic CO_2 storage projects. The mixed drive, which is a combination of two main mass transfer mechanisms (vaporizing and condensing drives), is the most common way in which CO_2 develops dynamic miscibility with reservoir oil. The minimum reservoir pressure at which dynamic miscibility between injected CO_2 and the reservoir can be achieved is termed as MMP. Though, industry's standard "slimtube test" is out there for longtime, the vanishing interfacial tension (VIT) technique and the Fast Fluorescence-Based Microfluidic MMP techniques appear to provide fast, cost-effective, and reliable MMP estimates, thus, becoming desirable design tools for future simultaneous CO_2-EOR and storage projects.

5.1 Introduction

At storage sites of currently operational LSIPs and other large-scale simultaneous CO_2-EOR and storage projects (Table 2.1), CO_2 injection (WAG or continuous CO_2 injection) is being done in miscible mode. However, condition of complete miscibility will occur only at a certain pressure, referred as minimum miscibility pressure (MMP). At pressures, lower than MMP, CO_2 and oil may remain immiscible or they just form a partially miscible mixture. The MMP is an important design parameter for assessing the achievement of complete mixing (i.e., miscibility) between injected CO_2 and reservoir oil irrespective their relative propositions at the time of mixing at a given location (from injection wellbore to the production wellbore). A reliable estimation of MMP is one of many initial tasks that are performed while designing a simultaneous CO_2-EOR and storage project as an overestimation of MMP can lead to increased operational or facility costs and an

© The Author(s) 2017   41
D. Saini, *Engineering Aspects of Geologic CO$_2$ Storage*, SpringerBriefs
in Petroleum Geoscience & Engineering, DOI 10.1007/978-3-319-56074-8_5

underestimation may result in less than expected oil recovery due to achievement of partial miscibility instead of complete miscibility.

5.2 CO$_2$-Oil Miscibility

Miscibility is a physical condition between two or more fluids under which they mix together in all proportions and all mixtures remain in a single fluid phase. For example, whether a mixture contains 90% (volume or mole) CO$_2$ and 10% (volume or mole) oil or 10% (volume or mole) CO$_2$ and 90% (volume or mole) oil, or any other mixing proportion for that matter, they will still completely mix to form a homogenous single phase. Miscibility is distinctly different from solubility. In case of solubility, when two or more phases are mixed together, formation of a single homogeneous phase depends on the mixing proportions of individual phases.

The formation of single fluid phase (i.e., achievement of miscibility condition) also implies that the solvent such as injected CO$_2$ will eventually displace all trapped oil from the pore that it invades. The development of miscibility between injected CO$_2$ and reservoir oil is an important factor in CO$_2$-EOR projects as it directly influences the residual oil saturation in the swept areas of the reservoir, thus, enhancing the microscopic displacement of oil to almost 100%.

At reservoir conditions of pressure and temperature greater than the critical pressure and temperature of CO$_2$ [7.4 MPa (1068 psi) and 31.1 °C (88 °F)], CO$_2$ behaves as a supercritical fluid (i.e., exhibiting both liquid-like and gas-like behaviors). These desired pressure and temperature conditions are normally encountered at reservoir depths greater than 2600 ft. The liquid-like behavior (high density) results in high absorption capacity because solubility increases with density, pressure, and temperature, whereas, gas-like behavior (high diffusivity and low viscosity) promotes high mass transfer rate between the solute and solvent. The lower critical (pressure and temperature) parameters of CO$_2$ allow the tuning of its solvent power at a low energy cost, thus, making it a solvent like no other.

5.3 Miscibility Mechanisms

When injected CO$_2$ meets the reservoir oil, a continuous mass transfer between the CO$_2$ and the oil occurs. As, the oil rich CO$_2$ phase and/or CO$_2$ rich oil phase move further away from the injection wellbore, they meet increasingly fresh oil to develop a bank of miscible phase which is produced at the production well. Injected CO$_2$ not only extracts some of the components out of the reservoir oil but some of the injected CO$_2$ also enters the oil phase. This combination of these two mass transfer mechanisms (mixed or vaporizing-condensing drive) is the most common way in which CO$_2$ develops the miscibility with reservoir oil (Fig. 5.1). In some cases, one mass transfer mechanism (i.e., extraction of the components from the oil

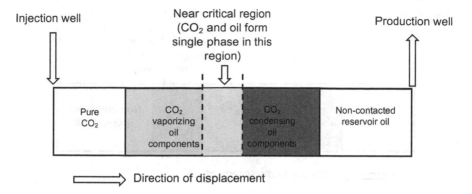

Fig. 5.1 One dimensional schematic of mixed (vaporizing + condensing) drive mechanism responsible for the development of CO_2-oil miscibility in the reservoir

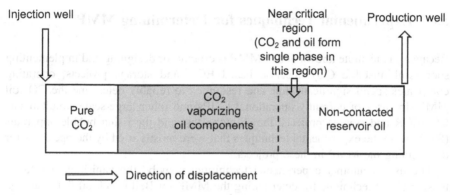

Fig. 5.2 One dimensional schematic of vaporizing drive mechanism responsible for the development of CO_2-oil miscibility in the reservoir

phase or vaporizing drive mechanism (Fig. 5.2) may dominate another mass transfer mechanism (i.e., entering of CO_2 into the oil phase or condensing drive mechanism (Fig. 5.3).

However, the miscibility condition described above is achieved via multiple/repeated contacts between injected CO_2 and reservoir oil, hence, it is referred as multiple contact or dynamic miscibility. If injected CO_2 and reservoir oil achieve miscibility condition without the need of multiple contacts, it is called first contact miscibility. The first contact miscibility is the one-time affair in real world. Injected CO_2 and reservoir oil must come in contact for multiple times as it moves away from the injection well to deep into the reservoir. Hence, for all practical purposes, it is the multiple or dynamic miscibility that is sought in the field. The minimum reservoir pressure at which dynamic miscibility between injected CO_2 and the reservoir can be achieved is termed as MMP.

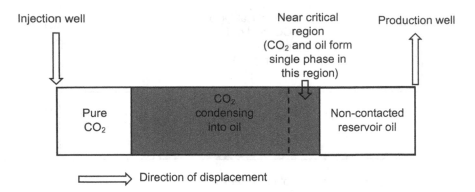

Fig. 5.3 One dimensional schematic of condensing drive mechanism responsible for the development of CO_2-oil miscibility in the reservoir

5.4 Experimental Techniques for Determining MMP

Because, an accurate knowledge of MMP is crucial for designing and implementing successful miscible CO_2 injection based EOR and storage projects, operating companies spend significant time and resources to reliably determine the CO_2-oil MMP. In cases of currently operational LSIPs and other large-scale simultaneous CO_2-EOR and storage projects, the slimtube test and the rising bubble apparatus (RBA) were the experimental techniques that were mostly used by the operators for determining the MMP in these projects.

The use of another experimental technique, namely, the vanishing interfacial tension (VIT) technique for determining the MMP for Bell Creek oil field (storage site for the Lost Cabin LSIP) has also been reported in the published literature (Gorecki 2015). In view of the time consuming and expensive nature of the experimental technique like slimtube test, the VIT technique appears to be a fast and expensive experimental way for providing the much-needed reliable estimate of MMP for simultaneous CO_2-EOR and storage projects.

Recently, Nguyen et al. (2015) has reported the use of a microfluidic approach, namely, Fluorescence-Based Microfluidic Method to quantify the CO_2-oil MMP at reservoir-relevant temperature. This technique leverages the inherent fluorescence of crude oils, is faster than conventional technologies, and provides quantitative, operator-independent measurements. In this method, mixing of continuously generated CO_2 bubbles within a microchannel embedded in silicon is observed quantitatively, leveraging the inherent fluorescence of crude oil sample. As emphasized by Nguyen et al. (2015), MMP can be obtained within 30 min using their microfluidic technique.

Though, industry's standard "slimtube test" is out there for longtime, alternative fast and cost-effective experimental techniques will be a desirable tool for future simultaneous CO_2-EOR and storage projects. In view of that, a brief discussion on both the VIT and the Fast Fluorescence-Based Microfluidic MMP techniques is provided next.

5.5 The VIT Technique

Rao (1997) developed a new experimental technique namely the vanishing inter-facial tension (VIT) technique that relies on studying the pressure dependence of the IFT behavior of CO_2 + crude oil system at several pressures steps and by making a plot of the IFT against the pressure. The MMP is then obtained by extrapolating the observed pressure versus IFT curve to zero IFT (Fig. 5.4).

The VIT technique reported by Rao and coworkers (Ayirala 2005; Ayirala and Rao 2006; Sequeira 2006; Sequeira et al. 2008; Saini and Rao 2010) uses a combination of the pendant drop (or falling-drop) technique and the capillary rise technique for experimentally studying the pressure dependence of IFT behaviors of complex CO_2-oil systems. Recently, Saini (2016) has investigated the robustness of both physical and numerical vanishing interfacial tension (VIT) experimentations in determining CO_2-oil MMP. He concluded that the physical VIT experimentation method in which the IFT measurements are made at varying pressures but with the same initial load of live oil and gas phases in the optical cell seems to be most

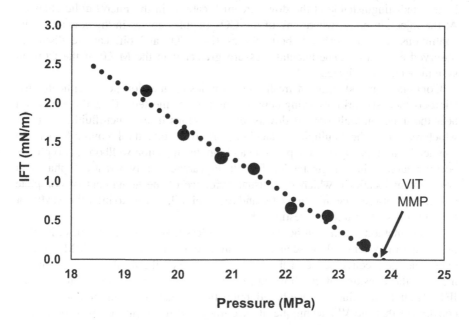

Fig. 5.4 Extrapolation of measured IFT versus pressure data to zero IFT for determining VIT MMP

robust procedure for experimentally studying the pressure dependence of IFT behaviors of complex CO$_2$-oil systems and thus determining the MMP using the VIT technique.

Ahmad et al. (2016) have also tested the accuracy and reliability of the VIT technique by comparing VIT determined MMPs, for same three considered crude oil samples, with slim tube determined MMPs. The concluded that since VIT estimated MMPs are in close match with more liable slim tube design determined MMPs, this technique can be utilized as a reliable and cheap alternative compared to more expensive and time consuming slim tube technique for accurate MMP determination without any potential of significant error.

5.6 Fluorescence-Based Microfluidic Method

In the fluorescence-based microfluidic method Nguyen et al. (2015), mixing of continuously generated CO$_2$ bubbles within a microchannel embedded in silicon, kept at reservoir-relevant temperature, is observed quantitatively. The inherent fluorescence of crude oil sample is used to visualize the deformation and ultimate disappearance of CO$_2$ bubbles introduced in the flowing oil stream in the microfluidic channel. As the pressure is increased, due to decrease in CO$_2$-oil IFT, injected CO$_2$ bubbles deform readily in response to flow induced stresses. As system pressure approaches the MMP, mixing is rapid, and the two phases are largely indistinguishable at the downstream locations in the microfluidic channel. At the injection point, where any of the CO$_2$ bubbles comes in the contact of the continuous flow of fresh oil, both phases (i.e., CO$_2$ and oil) can be distinctly observed even at an experimental pressure greater than the MMP of the CO$_2$-oil system being investigated.

Continuous introduction of fresh CO$_2$ bubbles at injection point replicates the "first-contact" scenario occurring between the freshly injected CO$_2$ and the fresh oil near the injection wellbore. At downstream locations in the microfluidic channel, which represents the "multiple-contact" scenario (CO$_2$ mixed oil moving forward to contact fresh oil occupying the pores away from the injection wellbore), two phases become largely indistinguishable. The disappearance of two distinct phases at downstream locations, which is a visual evidence of the attainment of complete miscibility between the injection CO$_2$ and reservoir oil, is used to infer the MMP for the CO$_2$-oil system under investigation.

In this method, MMP can be determined in less than an hour. Because, it relies on change in system's fluorescence for quantitative measurement of MMP, it also overcome the inconvenience of the VIT technique i.e., the need to experimentally measure the densities of equilibrated liquid and vapor phases for calculating the IFT. Though, the fluorescence-based microfluidic method is faster and is more convenient that the VIT technique, its accuracy and reliability needs to be further evaluated for the CO_2-oil systems that may exhibit condensing or mixed (condensing + vaporizing) drive mechanisms for attaining complete multiple-contact miscibility.

5.7 Other MMP Determination Techniques

Apart from above mentioned experimental methods, several other experimental, empirical, theoretical, and numerical methods are also available for determining MMP for CO_2-oil systems. These include various empirical correlations, rising bubble apparatus (RBA), micro slim-tube test, PVT multi-contact experiments, key tie-line approach and method of characteristics (MOC), vanishing tie-line approach, response surface based model, and linear gradient theory (LGT) model.

Various empirical correlations are great for performing initial project scoping studies as there may be limited information available on the PVT (pressure, volume, and temperature) properties of the reservoir under evaluation. If extensive PVT data are available for the reservoir oil, numerical methods can also be used to predicting the MMP.

References

Ahmad W et al (2016) Uniqueness, repeatability analysis and comparative evaluation of experimentally determined MMPs. J Petro Sci Eng 147:218–227

Ayirala SC (2005) Measurement and modeling of fluid-fluid miscibility in multicomponent hydrocarbon systems. Ph.D. Dissertation, Louisiana State University, Baton Rouge, Louisiana

Ayirala SC, Rao DN (2006) Comparative evaluation of a new MMP determination technique. Soc Petro Eng. doi:10.2118/99606-MS

Gorecki CD (2015) The plains CO_2 reduction (PCOR) partnership: Bell Creek field project. Carbon storage R&D project review meeting. Available via https://www.netl.doe.gov/File%20Library/Events/2015/carbon%20storage/proceedings/08-19_05_Gorecki.pdf. Accessed 25 Jan 2017

Nguyen P et al (2015) Fast fluorescence-based microfluidic method for measuring minimum miscibility pressure of CO_2 in crude oils. Anal Chem 87(6):3160–3164

Rao DN (1997) A new technique of vanishing interfacial tension for miscibility determination. Fluid Phase Equilib 139(1–2):311–324

Saini D (2016) An investigation of the robustness of physical and numerical vanishing interfacial tension experimentation in determining CO_2 + crude oil minimum miscibility pressure. J Petro Eng 2016:13. doi:10.1155/2016/8150752

Saini D, Rao DN (2010) Experimental determination of minimum miscibility pressure (MMP) by gas/oil IFT measurements for a gas injection EOR project. Soc Pet Eng. doi:10.2118/132389-MS

Sequeira DS (2006) Compositional effects on gas-oil interfacial tension and miscibility at reservoir conditions. MS Thesis, Louisiana State University, Baton Rouge, Louisiana

Sequeira DS, Ayirala SC, Rao DN (2008) Reservoir condition measurements of compositional effects on gas-oil interfacial tension and miscibility. Soc Pet Eng. doi:10.2118/113333-MS

Chapter 6
Maintaining the Integrity of Storage Sites

Abstract Maintaining the integrity of overlying caprock(s) and reservoir rock as well as the integrity of wellbores penetrating the caprock(s) and reservoir rock(s) is essential for long-term and safe storage of injected CO_2. The presence of a normally pressured caprock and maintaining a reservoir pressure below 50% of lithostatic pressure appears to be a good strategy for maintaining the integrity of caprocks in engineered CO_2 containment. Conservative approaches such as that the reservoir pressure does not exceed greatly the initial reservoir pore pressure and the CO_2 injection pressure remains below the maximum sustainable overpressure can help in minimizing the risk of irreversible mechanical damages to the reservoir rocks. Periodic integrity (pressure testing) of the well, leak testing of wellbore seals (packers, wellhead, and Xmas Tree) as well as chemical analysis of any fluids sampled from well annuli can also assist in maintaining the well integrity.

6.1 Introduction

The reservoir rock (injection zone) in which CO_2 is injected, injection and production wellbores that facilitate the injection of CO_2 and production of reservoir fluids (oil, gas, water) and part of the injected CO_2, and a seal (or a set of seals) to retain the injected CO_2 in the injection zone are the three basic elements of a storage sites. Apart from injecting CO_2 in miscible mode, maintaining the integrity of wellbores, injection zone(s), and the overlying seal(s) (caprock(s) is another important engineering aspect that needs to be carefully considered while designing a simultaneous CO_2-EOR and storage project. Before, an injection site can be established as a safe storage site, its ability to confine (i.e., no spill out of the reservoir) the injected CO_2 should also be reconfirmed.

The integrity of storage sites refers to geomechanical properties including *in situ* horizontal stresses, rock strength (tensile and compressive), stiffness properties [Poisson's ratio and elastic (Young's modulus)], and rock time-dependent deformation properties such as swelling potential, and dynamic properties (compressional wave velocities, shear wave velocity, dynamic Poisson's ratio, and dynamic

© The Author(s) 2017

D. Saini, *Engineering Aspects of Geologic CO2 Storage*, SpringerBriefs
in Petroleum Geoscience & Engineering, DOI 10.1007/978-3-319-56074-8_6

modulus). The geomechanical properties depend on lithology, pre-existing planes of weakness, regional geomechanical stresses, induced stress resulting from the reservoir fluid withdrawals and external fluids (water, gas and/or CO_2), and coupled geomechanical-chemical processes.

Because of certain portions of geologic formations have held hydrocarbons (oil and/or gas) for millions of years before being exploited, they are preferred as storage sites for geologic CO_2 storage projects. The extensive production and injection histories of numerous depleted oil fields without any incident of sudden migration of reservoir or injected fluids, except instances of naturally occurring seepage, provide additional evidence on their ability to safely store large volumes of reservoir and injected (water, CO_2, hydrocarbon gases etc.).

However, unlike in the commercial CO_2-EOR operations, in which systematic monitoring of the leakage (i.e., spill out of the reservoir) of injected fluids (most often water and/or CO_2) is not often required or minimally required, simultaneous CO_2-EOR and storage projects require mandatory monitoring of any potential spill out of the reservoir for ensuring the safe storage of injected CO_2 over long time spans (hundreds to thousands of years) at a storage sites. For effective monitoring, robust site characterization involving detailed assessments of the integrity of overlying caprock(s) and reservoir rock as well as the integrity of wellbores penetrating the caprock(s) and reservoir rock(s), is must.

6.2 Maintaining the Integrity of Overlying Caprock(s)

The presence of a top seal is one of the mandatory criteria that must be fulfilled while screening and selecting a depleted oil field as a storage site for a simultaneous CO_2-EOR and storage project. The buoyancy forces resulting from the density difference between injected CO_2 and reservoir fluids (oil and/or water) cause CO_2 to rise to the top of the injection zone (reservoir rock) and spread laterally. An effective top seal (caprock) will ensure that the injected CO_2 do not leak into other overlying strata or end up back into the atmosphere.

Impermeable formations, such as evaporates, shales and siltstones, tend to be finer grained or of a mixed grain size, with smaller, fewer, or less interconnected pores (IEAGHG 2008), making them an effective caprock. The exposure of a low-permeability caprock to CO_2-rich fluid over long period (thousands of years), is likely to affect only the lowest few meters (next to the reservoir rock) of the caprock with a small effect on hydraulic and mechanical properties (Rutqvist 2012). Also, most of the important coupled chemical-mechanical effects are expected to occur along the faults and fractures whether naturally occurring or created by high CO_2 injection pressure (Rutqvist 2012). However, impact of such chemical-mechanical effects would have a minor impact on shales or mudstone caprocks (Rutqvist 2012).

According to study published by the International Energy Agency's (IEA) GHG program (IEAGHG 2008), to form an effective seal (i.e., to prevent vertical migration of injected CO_2 and/or any of the reservoir fluids out of the reservoir), the sealing lithology needs to be impermeable to CO_2, upfaulted and relatively ductile (resistance to fracturing), and laterally continuous while maintaining a consistency of properties over a large area. A detailed treatment of the topic of caprock systems for geologic CO_2 storage can be found it another report (IEAGHG 2011) published by the IEAGHG. Song and Zhang (2012) have also provided a comprehensive review of caprock-sealing mechanisms that are critical for geologic CO_2 storage projects.

The presence of a thick caprock or several layers of impermeable rocks increases the confidence in its much larger and continuous areal extent compared to relatively thin (i.e., few ft. thick caprock) single caprock. Also, it reduces the possibility of leakage due to capillary (i.e., scenario of capillary entry pressure of the caprock being larger than the buoyance pressure of the stored CO_2) or molecular diffusion effects. Reliable estimation of the thickness of overlying caprock(s), which can be estimated from well logs, drill cuttings, and stratigraphic calculations, is one of the key activities that are carried out in initial site characterization of the potential storage site.

The caprock capillary entry pressure is a function of pore size distribution of caprock, the wetting characteristics of caprock/reservoir fluids/CO_2 system, the density of the injected CO_2 and reservoir fluids (oil, gas, water) and the IFT of CO_2/ reservoir fluids. Laboratory evaluations of capillary entry pressure and molecular diffusion effects, often rare, further ensures the sealing quality of the caprock. Li et al. (2005) has reported on such laboratory testing performed for Midale Evaporite, the low-permeability overlying seal for the Weyburn storage site. Wollenweber et al. (2010) provides an excellent experimental treatment of capillary and molecular diffusion effects on CO_2 sealing efficiency of caprocks.

Another key characteristic that makes a caprock an effective caprock(s) is the absence of pre-existing natural fractures and faults. The available seismic surveys, well logs or cores, or outcrop analogs are normally used for verifying the absence of faults in the overlying caprock(s). However, some faults may remain undetected if they fall below the resolution limits of these techniques. In case of depleted oil fields, significant reservoir pore pressure changes (decrease during historical oil and gas production phase and increase during waterflooding and/or CO_2 injection phase) can cause induce slip on faults and hydromechanically fracture or reactivate the caprock. The hydromechanically induced/reactivated fractures or faults can provide pathways for CO_2 leakage out the reservoir and can potentially compromise the integrity of the overlying caprock(s). The results of studies (Rohmer and Bouc 2010 in Song and Zhang 2012) on hydromechanical failure of caprock indicate that the initial *in situ* stress state is the most sensitive parameter and that storage sites with the lowest initial *in situ* stress state present the highest risk of caprock failure. The Poisson's ratios of caprock and reservoir rocks also have big influence on the potential hydromechanical failure of caprocks.

If natural fractures and faults are present in the caprock, it is ensured that they are sealed (closed) and CO_2 injection would not cause any significant failure (re-opening) of them. To avoid any potential failure of faults and fractures present near the injection zone, CO_2 injection pressure is carefully selected and pressurization is also carefully monitored via customized MVA/MMV program (geophysical, geochemical, and atmospheric monitoring) as leakage through faults or fracture networks can be rapid and catastrophic (Song and Zhang 2012).

The Weyburn-Midale LSIP is one of the operational LSIPs, for which the results of comprehensive geomechanical and geochemical studies carried out for ensuring the safe storage of injected CO_2 are easily available in public domain (Verdon et al. 2013). Simultaneous injection and production during CO_2-EOR operations at Weyburn has caused reservoir pore variations across the field causing geomechanical deformation resulting in microseismic events. According to Verdon et al. (2013), all the recorded microseismicity is located within 198 m (650 ft.) of the top of the reservoir, implying that the induced deformation has not created pathways for fluid flow beyond the containment complex. The induced deformation in the overburden appears to occur due to stress transfer resulting from lower pore pressures at the production wells rather than a hydraulic connection. This implies that the current deformation does not pose a direct risk to storage security. However small discontinuities are present in the overburden that are close to failure criteria. The presence of larger faults near top of reservoir, which may have been close to failure as well, could present an increased risk to caprock integrity as well as it can lead to seismic events that can be felt (Verdon et al. 2013).

Single incident of fault reactivation may not necessarily open a new flow path for leakage, thus compromising the caprock integrity (Rinaldi et al. 2014). It means a single event is generally not enough to substantially change the permeability along the entire fault length and even if some changes in permeability occur, this does not mean that the CO_2 will migrate up along the entire fault, breaking through the caprock to enter the overlying aquifer (Rinaldi et al. 2014).

An understanding of the mechanisms for CO_2 leakage in natural CO_2 reservoirs and gas storage reservoirs, which can be considered a good natural analog for geologic CO_2 storage sites, can further assist in developing a better strategy for maintaining the integrity of caprock. Petroleum industry has sufficient experience and expertise in managing them. Based on an analysis of global dataset of 49 natural CO_2 reservoirs (Miocic et al. 2013), the presence of a normally pressured (pore pressure equal to normal hydrostatic pressure) caprock and maintaining a reservoir pressure below 50% of lithostatic (geostatic or overburden) pressure, which is slightly above the hydrostatic pressure encountered in normally pressured oil fields, appears to be a good strategy for maintaining the integrity of caprocks in engineered CO_2 containment (simultaneous CO_2-EOR and storage).

Bruno et al. (2014), have established a set of parameters (risk factors) that influences the likelihood of caprock failure. They have also established the magnitude value ranges for each parameter, which, when applied to particular geologic and operation settings, can quantify the risk. The quantitative risk and decision analysis tools such as developed by Bruno et al. (2014), can provide a mean to rapid assess the potential for leakage during CO_2 injection and to compare potential and active storage sites.

6.3 Maintaining the Integrity of Reservoir Rocks

The Changes in the reservoir pore pressure and volume of the rock occurring due to depletion and repressurization of reservoir rocks, may, occasionally, results in formation deformation in oil fields causing detectable subsidence and/or uplift of the earth surface as well as induced seismic activity. Subsidence of the earth surface because of pressure changes induced by hydrocarbon extraction has been reported in oil fields such as Wilmington oil field (Pierce 1970), a prolific sandstone reservoir in California, USA, the Ekofisk oil field (Nagel 2001), located in the southern part of the North Sea, Norway, producing from naturally fractured chalk reservoirs.

Frequent incidents of induced seismicity have clearly been established only for two hydrocarbons bearing sedimentary basins, namely the Permian basin, Texas, and Rotliegendes, Netherlands, however, even these two regions not all hydrocarbon fields are equally affected (Suckale 2009). The induced seismicity in the Netherlands is probably primarily the result of reactivation of normal faults in the reservoir, whereas, seismicity in the Permian basin appears to be correlated to reservoir production, fluid migrations, natural occurrence of overpressured fluids, tectonic activity, and EOR operations (Suckale 2009). The Permian basin is the home for numerous commercial CO_2-EOR projects as well as three operational LSIPs, namely, Val Verde, Air Products, and Century Plant (Table 1.1).

Both fluid injection and fluid extraction are typically associated with induced seismicity in hydrocarbon fields, however, several events with moderate to large seismic magnitudes have also been related to fluid injections (Suckale 2009). Subsidence can also be related to seismicity as evident from moderate event at Ekofisk oil field and major event at Wilmington oil field, however, in some cases such as Belridge oil field, California, USA, it may be largely aseismic (Suckale 2009).

When fluids are injected into the reservoir rock, an increase in reservoir pore pressure may result in an imbalance of in situ stresses or activation of a fault, which is detrimental for reservoir rock integrity. Similarly, one of the consequences of CO_2 injection in depleted oil fields is the alternation in the *in situ* thermal and pressure stresses that have potential to impact the mechanical properties of the reservoir rock(s). An alternation in the *in situ* stresses (vertical or normal stress and horizontal stresses) resulting from changes in reservoir pore pressure and volume of the reservoir rock can lead to the loss of reservoir and caprock integrity, and the re-activation of existing faults (Orlic and Schroot 2005). The formation of carbonic acid resulting from dissolution of injected CO_2 in reservoir water and its reaction with carbonate reservoir rocks can also result in increased porosity leading to chemical compaction, thus affecting the mechanical properties such as rock stiffness and strength. High porosity carbonate rocks are, often, more prone to occurrence of chemical compaction.

The reservoir pressurization due to CO_2 injection can also cause vertical expansion of the reservoir which may result in a detectable uplift of ground surface (e.g., In Salah dedicated CO_2 storage project in Algeria), however, the magnitude of uplift will depend on the geometry, geomechanical properties (such as compressibility) of the reservoir and surrounding sediments, and the thickness of the underground reservoir being pressurized at depth (Rutqvist 2012).

During injection, the vertical stress relatively remains constant (equal to the weight of overburden), however, depending on the geometry and geochemical properties (such as compressibility) of the reservoir-caprock system, horizontal stress will change with the injection (Rutqvist 2012). A change in horizontal stress is a source of potential mechanical inelastic responses, including shear reactivation of existing fractures that could result in seismic events (Rutqvist 2012). At Weyburn LSIP, during the monitoring of injection-induced microseismicity, less than 100 events with magnitudes ranging from -1 to -3 have been recorded, documenting low rate of seismicity (Rutqvist 2012). As discussed earlier, many events recorded at Weyburn were in the overburden outside the injection zone and interpreted to be triggered by stress transfer from injection-induced expansion of the reservoir (Verdon et al. 2013).

If we look at the pressure depletion and repressurization histories of some of the storage sites of currently operational LSIPs, local land subsidence in the Hastings Field area, storage site (West Hastings oil field) for one of the operational LSIPs (Air Product), has also been reported in the past (Holzer and Bluntzer 1984). According to Holzer and Bluntzer (1984), although ground-water withdrawal is undoubtedly the most important factor contributing to the subsidence, oil and gas withdrawal may be partially responsible for the differential subsidence.

The West Hastings storage site has hydrologically isolated fault blocks (Porse 2013). It also has a main growth fault extending to the surface and several cross faults and production history suggests that these cross faults maybe somewhat cross-fault transmissive (Hovorka 2010 in Mehlman 2010). However, the presence of large commercially exploitable hydrocarbon accumulation and the historical records of isolation between zones in both sides of the faults, it is assumed that, for pressures below the original reservoir pressure, these faults will act as a seal (Ortega 2012). For facilitating a miscible CO_2 flooding and for minimizing the losses of injected CO_2 into the aquifer, operator has begun downdip water injection (re-pressurization) in the reservoir (Davis et al. 2011a, b). The updip bounding faults have acted as no flow boundary thus allowing for improved re-pressurization (Davis et al. 2011a, b).

The Bell Creek oil field, storage site for another currently operational LSIP (Lost Cabin), though discovered as subnormally pressured (i.e., initial reservoir pressure significantly lower than normal hydrostatic pressure), appears to be a hydrodynamically isolated system with effective reservoir seals (Hamling 2013). The reservoir has undergone pressure depletion and significant re-pressurization by water injection before becoming the storage site for Lost Cabin LSIP (Burt et al. 1975; Welch 2012).

The dedicated MVA/MMV programs at both West Hastings and Bell Creek storage sites include monitoring technologies (Table 3.1) that are designed to detect any stress-induced seismicity and out of zone vertical migration of injected CO_2 for ensuring the integrity of reservoir-caprock systems. However, for minimize the risk of irreversible mechanical damages, conservative approaches that can be taken are that reservoir pressure does not exceed greatly the initial reservoir pore pressure (Kovscek 2002 in Li 2016) and the CO_2 injection pressure remains below the maximum sustainable overpressure (Varre et al. 2015 in Li 2016).

6.4 Maintaining the Integrity of Wellbores

Well integrity means the achievement of fluid containment and pressure containment within the well throughout its whole life cycle (Smith et al. 2011). Maintaining the wellbore integrity is critical for the success of simultaneous CO_2 EOR and storage projects because wellbores pose maximum risk of CO_2 migration from the reservoir. Not only, the existing and abandoned wells within the storage site area, that have already penetrated the primary seal (caprock), should be leak free, but also the CO_2 injection wells should also be designed for the long operational lives (often several decades) and even longer (exceeding hundreds to thousands of years) integrity after abandonment.

Wellbore integrity issues are usually divided into two types: improper completion and abandonment of the wells; and the long-term stability of wellbore materials in a CO_2-rich environment (Carey et al. 2009). Well integrity can be compromised by defective well completion or as a result of chemical and mechanical stresses that damage the well during the operation or abandonment phases (Carroll et al. 2016).

A wide range of wellbore completions and abandonments can be encountered in the hydrocarbon extraction and geologic storage projects. However, most often, it is the poor cementing job (i.e., poor mud displacement during cementing job, gas channeling through unset cement) that results in poor cement bonds at cement/reservoir rock, cement/caprock, and cement/casing interfaces, thus, allowing potential for CO_2 leakage through wellbore and corrosion of casing material(s). Stress-induced cracking, formation of microannuli at the casing-cement interface, incomplete cementing in the annular space or cement degradation can expose the casing to fluids including reservoir brine, injected CO_2, and associated impurities like H_2S and organic acids, thus leading to an increased likelihood of sustained casing pressure (SCP) and casing corrosion (Choi et al. 2013).

According to Carey et al. (2009), an increasing number of field studies, experimental studies, and theoretical considerations indicate that the most significant leakage mechanism is likely to be flow of CO_2 along the casing-cement microannulus, cement-cement fractures, or the cement-caprock interface. The magnitude of flows along these interfaces is a complex function of the pressure gradient, geomechanical properties that support the interface and dissolution/precipitation reactions that lead to widening or closure of the interface (Carey et al. 2009).

Geomechanical processes can also impact well integrity both during drilling and completion, as well as during the actual CO_2 injection (Rutqvist 2012). Drilling and completion of wells through layers of shale and mudstone requires special attention, because rock failure and deformation associated with wellbore instability may damage wellbores, leading to increased permeability and potential leakage paths (Orlic 2009 in Rutqvist 2012). Wellbore stresses will change over time, as a result of shifts in pressure due to pumping or leakage, fluctuations in subsurface temperature (also potentially a byproduct of injection or leakage), or due to natural variations in subsurface conditions (tectonic stresses and seismicity), thus affecting the wellbore integrity (Carroll et al. 2016).

At Weyburn-Midale, storage sites for Great Plains Synfuel and Boundary Dam LSIPs, integrity of abandoned wells, was identified as being a significant risk factor with respect to the permanence of injected CO_2 (Choi et al. 2013). Many out of over 4500 wells in the area where CO_2 injection is occurring in Weyburn and Midale Oil Units, are approaching 60 years old and have been completed with a varied of methods using different casing steels and various grades of cement (Choi et al. 2013). On another hand, many of the US CO_2-EOR projects have been in services since 1970s, there is not yet long-term experience of the abandonment (storage) phase of the project life to indicate how the well integrity is maintained over time (Smith et al. 2011). In an analysis and performance of 30 years of CO_2 exposure from the SACROC Unit, storage site of one of the operational LSIPs (Val Verde), casing corrosion was found minimal to non-existent, however, casing corrosion may also be a used as a tool detecting long-term flow along the casing-cement interface behind-casing fluid flow using relatively simple logging studies (Carey et al. 2009). At the West Hastings oil field, storage site for Air Products LSIP, CO_2 injection wells with relatively smaller wellbore casing diameters were identified as more prone to CO_2 leakage (Li 2014).

Thorough evaluation of the cement jobs is acritical for corrosion prediction and protection, as well as assurance of integrity of existing, abandoned, and new wells (Choi et al. 2013). However, the well completion records for old wells are hard to find. In such scenarios, sustained annulus pressure (SAP), that is, pressure within the annuli of the well which cannot be reduced to zero by bleeding off (Smith et al. 2011), can be used for identifying any developed leakage. Daily monitoring and intermittent testing of SAP can be used as an effecting monitoring and warning tool for early detection of well integrity issues. Periodic integrity (pressure testing) of the well, leak testing of wellbore seals (packers, wellhead, and Xmas Tree) as well as chemical analysis of any fluids sampled from well annuli can also assist in maintaining the well integrity.

6.5   Strategy for CO_2 Leakage Prevention and Remediation

A 2007 report on the remediation of leakage from CO_2 storage reservoirs published by the IEAGHG (2007) provides a comprehensive strategy for preventing and remediating any potential CO_2 leakage at a storage site. It has following five elements:

1. Selecting favorable storage sites with low risks of CO_2 leakage. Initial characterization of the potential storage sites will assist in identifying a safe and secure site.
2. Identification of all old abandoned wells in the vicinity of storage site, design and installation of injection wells so that they are resistant to CO_2, and proper closure of the storage site are the three key priorities for ensuring the well integrity.
3. Conduct a phased series of formation simulation-based modeling for tracking and predict the location and movement of the CO_2 plume. Modeling results should be calibrated and updated accordingly as more site specific geological

and reservoir data become available via drilling of injection and observation wells, and repeat surveys after injection.

4. Install and maintain a comprehensive monitoring system, which is designed as an early warning system of any impeding CO$_2$ leakage event, and to provide on-going information on the movement and immobilization of the CO$_2$ plume.

5. Establish a "Ready-to-Use" contingency plan/strategy, should a CO$_2$ leakage event occur. The plan should contain remediation options for all the most likely leakage scenarios.

Compared to the total cost of a geologic CO$_2$ storage project, cost associated with the implementations of above mentioned strategy for remediation of a potential CO$_2$ leakage event can be considered relatively low, unless a leakage event is of catastrophic nature. Also, use of a prudent approach for maintaining the integrity of the storage site will not only help the industry in increasing the public acceptance of geologic CO$_2$ storage, but, it will also ensure the public safety.

References

Bruno MS et al (2014) Development of improved caprock integrity analysis and risk assessment techniques. Energy Procedia 63:4708–4744

Burt RA, Haddenhorst FA, Hartford JC (1975) Review of Bell Creek waterflood performance—Powder River, Montana. Soc Pet Eng 27:1–443. doi:10.2118/5670-PA

Carey JW et al (2009) Wellbore integrity and CO$_2$-brine flow along the casing-cement microannulus. Energy Procedia 1(1):3609–3615

Carroll S et al (2016) Review: role of chemistry, mechanics, and transport on well integrity in CO$_2$ storage environments. Int J Greenhouse Gas Control 49:149–160

Choi Y-S et al (2013) Wellbore integrity and corrosion of carbon steel in CO$_2$ geologic storage environments: a literature review. Int J Greenhouse Gas Control 16S:S70–S77

Davis DW et al (2011a) Large scale CO$_2$ flood begins along Texas Gulf Coast. Society of petroleum engineers. doi:10.2118/144961-MS

Davis DW et al (2011b) Large scale CO$_2$ flood begins along Texas Gulf Coast: technical challenges in re-activating an old oil field. Annual CO$_2$ Flooding Conference, Midland, Texas. Available via http://www.co2conference.net/wp-content/uploads/2012/12/3.2-Denbury_Davis_Hastings_2011-CO2Flooding_Conf.pdf. Accessed 29 Jun 2016

Hamling JA (2013) Overview of the Bell Creek combined CO$_2$ storage and CO$_2$ enhanced oil recovery project. Energy Procedia 37:6402–6411

Holzer TL, Bluntzer RL (1984) Land subsidence near oil and gas fields, Houston, Texas. Ground Water 22:450–459. doi:10.1111/j.1745-6584.1984.tb01416.x

Hovorka SD (2010) Site-specific MVA options evaluation, Gulf Coast Carbon Center—Report to Denbury on MVA Planning for FOA 15. In: Mehlman S (2010) carbon capture and sequestration (via enhanced oil recovery) from a hydrogen production facility in an oil refinery. Available via http://www.osti.gov/scitech/servlets/purl/1014021. Accessed 25 Jun 2016

IEAGHG (2007) Remediation of leakage form CO$_2$ storage reservoirs. International Energy Agency (IEA) Greenhous Gas R&D Programme. 2007/11. Available via http://ieaghg.org/docs/General_Docs/Reports/2007-11_Remediation%20Report.pdf. Accessed 22 Feb 2017

IEAGHG (2008) Assessment of subsea ecosystems impacts. International Energy Agency (IEA) Greenhouse Gas R&D Programme. 2008/8. Available via http://hub.globalccsinstitute.

com/sites/default/files/publications/95761/assessment-sub-sea-ecosystem-impactspdf.pdf. Accessed 28 Jan 2017

IEAGHG (2011) Caprock systems for CO_2 geological storage. International Energy Agency (IEA) Greenhouse Gas R&D Programme. 2011/01. Available via http://www.ieaghg.org/docs/ General_Docs/Reports/2011-01.pdf. Accessed 28 Jan 2017

Kovscek AR (2002) Screening criteria for CO_2 storage in oil reservoirs. Pet Sci Technol 20:841–866

Li B (2014) A preliminary assessment of leakage possibility of CO_2 sequestration wells in two Gulf Coast fields. MS Thesis, University of Louisiana at Lafayette, Louisiana

Li G (2016) Numerical investigation of CO_2 storage in hydrocarbon field using a geomechanical-fluid coupling model. Petroleum 2(3):252–257

Li S et al (2005) Gas breakthrough pressure for hydrocarbon reservoir seal rocks: implications for the security of long-term CO_2 storage in the Weyburn field. Geofluids 5:326–334. doi:10.1111/ j.1468-8123.2005.00125.x

Mehlman S (2010) Carbon capture and sequestration (via enhanced oil recovery) from a hydrogen production facility in an oil refinery. Available via http://www.osti.gov/scitech/servlets/purl/ 1014021. Accessed 25 Jun 2016

Miocic JM et al (2013) Mechanisms for CO_2 leakage prevention—a global dataset of natural analogues. Energy Procedia 40:320–328

Nagel NB (2001) Compaction and subsidence issues within the petroleum industry: from Wilmington to Ekofisk and beyond. Phys Chem Earth Part A Solid Earth Geod 26:3–14

Orlic B (2009) Some geomechanical aspects of geological CO_2 sequestration. KSCE J Civ Eng 13:225–232

Orlic B, Schroot B (2005) The mechanical impact of CO_2 injection. European Association of Geoscientists and Engineers. 67th EAGE Conference and Exhibition, Feria de Madrid

Ortega CAP (2012) A value of information analysis of permeability data in a Carbon, Capture and Storage project. MS thesis. The University of Texas at Austin

Pierce RL (1970) Reducing land subsidence in the Wilmington oil field by use of saline waters. Water Resour Res 6(5):1505–1514. doi:10.1029/WR006i005p01505

Porse SL (2013) Using analytical and numerical modeling to assess deep groundwater monitoring parameters at carbon capture, utilization, and storage sites. MS Thesis. University of Texas at Austin, Texas

Rinaldi AP, Rutqvist J, Cappa F (2014) Geomechanical effects on CO_2 leakage through fault zones during large-scale underground injection. Int J Greenhouse Gas Control 20:117–131

Rohmer J, Bouc O (2010) A response surface methodology to address uncertainties in cap rock failure assessment for CO_2 geological storage in deep aquifers. Int J Greenhouse Gas Control 4 (2):198–208

Rutqvist J (2012) The geomechanics of CO_2 storage in deep sedimentary formations. Geotech Geol Eng 30:525. doi:10.1007/s10706-011-9491-0

Smith L et al (2011) Establishing and maintaining the integrity of wells used for sequestration of CO_2. Energy Procedia 4:5154–5161

Song J, Zhang D (2012) Comprehensive review of caprock-sealing mechanisms for geologic carbon sequestration. Environ Sci Technol 47(1):9–22. doi:10.1021/es301610p

Suckale J (2009) Induced seismicity in hydrocarbon fields. Adv Geophys 51:55–106 Elsevier

Varre SBK et al (2015) Influence of geochemical processes on the geomechanical response of the overburden due to CO_2 storage in saline aquifers. Int J Greenhouse Gas Control 42:138–156

Verdon JP et al (2013) A comparison of geomechanical deformation induced by 'megatonne' scale CO_2 storage at Sleipner, Weyburn and In Salah. Proc Natl Acad Sci 110:E2762–E2771. doi:10. 1073/pnas.1302156110

Welch R (2012) Bell Creek CO_2 development update. Available via https://www.uwyo.edu/eori/_ files/co2conference12/russ_denbury_bell%20creek%20wyoming%20co2%20conference_ 07122012_modified.pdf. Accessed 26 Jun 2016

Wollenweber J et al (2010) Experimental investigation of the CO_2 sealing efficiency of caprocks. Int J Greenhouse Gas Control 4(2):231–241

Chapter 7
Selection of Favorable Storage Sites

Abstract Majority of current geologic storage sites have largely been selected on the basis of petroleum industry's practical experience and efforts that are used for screening suitable candidates (depleted oil fields) for commercial CO_2-EOR projects. However, now, there are several methodologies and best practices are available for screening, selection, and characterization of favorable storage sites for future simultaneous CO_2-EOR and storage projects. These methodologies and best practices can be used to the candidate sites according to their CO_2-EOR and storage potential. After a regional proximity analysis and social context analysis for top candidate storage sites, initial characterization should be performed for most promising sites. Because detailed characterization is very expensive and lengthy process, hence, detailed characterization of any of the initially characterized sites should be con-ducted for only those sites that are being planned for commercial project status.

7.1 Introduction

The selection of safe and secure sites in the first place is the most important aspect for not only to increase the public acceptance of geologic CO_2 storage, but also, it is the most important condition for a technically and economically successful commercial simultaneous CO_2-EOR and storage projects. Yes, various engineering aspects discussed in previous chapters are equally important, however, if they are implemented at favorable storage sites, which are carefully screened, selected, and characterized, their benefits would increase by several folds.

So far, public (i.e., funding and research and development (R&D) support from the agencies like the DOE and IEA) and private (operational experience and expertise of field owner/operators and service companies) partnership has done a marvelous job in carefully selecting favorable storage sites, conducting additional site characterization, risk analysis and simulation, well management, MVA/MMV, and public outreach activities necessary to demonstrate the viability of geologic CO_2 storage. However, safe and secure storage of captured CO_2 at a storage site

© The Author(s) 2017
D. Saini, *Engineering Aspects of Geologic CO₂ Storage*, SpringerBriefs in Petroleum Geoscience & Engineering, DOI 10.1007/978-3-319-56074-8_7

ultimately rely on the process of identifying and fully characterizing of potential storage sites.

The selection of majority of currently operational LSIPs for simultaneous CO_2-EOR and storage projects had largely stemmed from petroleum industry's practical experience and efforts used for screening suitable candidates (depleted oil fields) for CO_2-EOR projects. For example, CO_2-EOR operations at SACROC Oil Unit were started in 1970s, well before geologic CO_2 storage become a known technology. Similarly, the Weyburn-Midale, West Hastings, Bell Creek, and several Pinnacle Reefs within Michigan's Northern Reef Trend were either already identified as top candidate for commercial CO_2-EOR and/or operators were already in the process of implementing CO_2-EOR in these depleted oil fields before the CCS technology was seriously considered, supported, and promoted by governmental and regulatory agencies and public and private funding agencies. The R&D efforts led by the governmental and public agencies such as National Energy Technology Laboratory's (NETL), IEA GHG Programme, have also helped the operators of these LSIPs in performing additional and necessary characterization activities for establishing these depleted oil fields as a reliable storage sites.

In the process, agencies like the NETL and the IEA GHG Programme have also established a framework and methodology for site screening, site selection, and initial characterization of a potential geologic CO_2 storage site. Such efforts include a compilation of best practices, communicate the experience grained through the U.S. Department of Energy's (DOE) Regional Carbon Sequestration Partnership (RCSP) initiatives, and develop a set of guidelines for communicating project related storage resources and risk estimates associated with a geologic CO_2 storage project. One of such Best Practice Manuals (BPMs) provides detailed guidelines for site screening, selection, and initial characterization (NETL 2013). However, let's first, take a look at the criteria that have been used by the petroleum industry for preliminary screening of the candidate oil fields for commercial CO_2-EOR projects.

7.2   Preliminary Screening of Oil Fields for Commercial CO_2-EOR

When it comes to select candidate oil fields for commercial CO_2-EOR projects, there are several technical screening criteria/guidelines that are reported in published literature (Taber et al. 1997; Meyer 2007; Adasani and Bai 2011; Moreno et al. 2014; Yin 2015). These criteria/guidelines are mainly built on the publicly available information on past and present EOR projects.

Moreno et al. (2014) have identified oil gravity (API), viscosity (cp), and depth (ft) as three most significant input parameters for miscible CO_2-EOR projects. Based on the analysis of field data collected from 134 CO_2-EOR projects in the U.S., Yin (2015) recommended following values of oil gravity, viscosity, and depth that should be looked for in a candidate oil field for selecting for miscible CO_2 flooding.

Sandstone reservoir:

oil gravity > 27 API
oil viscosity < 3 cp
Depth > 350 m (1150 ft).

Carbonate reservoir:

oil gravity > 28 API
oil viscosity < 6 cp
Depth > 914 m (3000 ft).

Yin (2015) also recommended that the oil saturation at the beginning of the CO_2 flood should be more that 20% pore volume (PV) and the favorable net pay thickness is from 22.8 m (75 ft) to 41.8 m (127 ft). Obviously, one can use these key reservoir parameters to perform a preliminary screening of candidate oil fields. However, initial screening of candidate storage sites for simultaneous CO_2-EOR and storage projects involves a different approach.

7.3   Preliminary Screening of Oil Fields for Simultaneous CO_2-EOR and Storage

First, the formations at depths greater than 800 m (2600 ft.) are considered best injection zones for simultaneous CO_2-EOR and storage projects as CO_2 will be in a supercritical state [7.4 MPa (1071 psi) and 31.1 °C (88 °F)] at these depths, thus, it will be effectively stored. Second, the potential injection zones should be below the deepest underground sources of drinking water (USDWs). Third, overlain confining zone should be comprised of one or more thick and impermeable confining intervals of sufficient lateral extent. If depleted oil fields present in a hydrocarbon producing basin or sub-basin meets all three criteria mentioned above, oil gravity and viscosity along with a preliminary estimated MMP using an empirical correlation based procedure (ARI 2005) may be used for narrowing down the list of candidate oil fields.

The publicly available geology and reservoir data from the sources such as published and open-file reports, atlases, and databases by state geological surveys, departments of natural resources, and academic institutions or data acquired from private firms can be used for performing this preliminary screening step.

Next, the ability to store desired volume of captured CO_2 at a given injection capacity (million tonnes per year) for a given project duration should be estimated for the candidate oil fields. Recently published studies such as Godec et al. (2011), Saini (2015) describe methodologies that can be used for estimating the CO_2 storage capacity of the top candidate oil fields.

Based on the collected data and interpretation, a list of top candidate depleted oil fields can be prepared and they can be ranked according to their CO_2-EOR and storage potential. Now, for the selected storage sites, a regional proximity analysis and social context analysis (NETL 2013) should be performed.

7.4 Regional Proximity and Social Context Analyses

Proximity between the source (stationary CO_2 emission source) and the sink (geologic storage site(s)) is the first step of the recommended regional proximity analysis. Apart from this initial step, routing, availability, use, and any conflict (e.g., right-of-ways) of the existing and potentially new pipelines and gathering lines/systems should be studied. Also, the existing resource development such as the wells that penetrate the confining zone, potential for conflicts between proposed storage project(s) and existing or prospective mineral leases as well the availability of complementary or competing infrastructure should be considered.

Next, data on the population centers, identification of any protected and sensitive areas, and evaluation of potential for other surface sensitivities (e.g., flood, landslide, wildfire, tsunami) in the proximity of the storage site(s) and their potential for any conflicts with siting of pipeline routes and another infrastructure need to be analyzed. Demographic and land use trends, public perception local economic and industrial trends, and community sensitivities to land use and the environment are the social context analyses that need to be performed. The data for these analyses can be found through governmental and commercial sources.

7.5 Selection of Qualified Storage Site(s) and Initial Characterization

The NETL BPM (NETL 2013) provides detailed guidelines for thorough evaluation of selected qualified storage site(s). The process includes site selection and initial characterization of the storage site(s) found qualified in initial site screening. The site selection step resembles to the petroleum industry's exploration approach of identifying "Exploration Play Areas" in a "Prospective Play". The top ranked qualified site(s) determined via site selection step are then assessed in greater detail during initial characterization step. The initial characterization process is analogous to processes that petroleum industry uses for "Prospect" evaluation during exploration phase. Once initial characterization is completed, qualified site is elevated to site characterization phase.

Both site selection and initial characterization steps are time consuming and expensive due to the demand of rigorousness involved in these steps. However, as can be seen in the case of currently operational North American LSIPs, operators had already wealth of the geologic and reservoir characterization data for the storage sites such Weyburn and Midale Oil Units, SACROC Oil Unit, West Hastings, and Bell Creek which helped all of the stakeholders (e.g., operator, DOE, and RCSP) in moving forward with site selection and initial characterization steps swiftly and completing these steps in cost effective manner. This is another reason that depleted oil fields are considered favorable storage sites.

Once, initial characterization for qualified sites is completed. Qualified sites are ranked for performing detailed characterization. Because detailed characterization is very expensive and lengthy process, hence, detailed characterization of any of the initially characterized sites should be conducted for only those sites that are being planned for commercial project status.

References

Adasani AA, Bai B (2011) Analysis of EOR projects and updated screening criteria. J Petro Sci Eng 79:10–24

ARI (2005) Basin oriented strategies for CO_2 enhanced oil recovery: onshore California oil basins. U.S. Department of Energy. Available via http://www.adv-res.com/pdf/Basin%20Oriented% 20Strategies%20-%20California.pdf. Accessed 11 Dec 2013

Godec M et al (2011) CO_2 storage in depleted oil fields: the worldwide potential for carbon dioxide enhanced oil recovery. Energy Procedia 4:2162–2169

Meyer JP (2007) Summary of carbon dioxide enhanced oil recovery (CO_2EOR) injection well technology. American Petroleum Institute. Available via http://www.api.org/ ~ /media/files/ ehs/climate-change/summary-carbon-dioxide-enhanced-oil-recovery-well-tech.pdf. Accessed 22 Jan 2017

Moreno JE et al (2014) EOR advisor system: a comprehensive approach to EOR selection. Int Pet Technol Conf. doi:10.2523/IPTC-17798-MS

NETL (2013) Best practices: site screening, selection, and initial characterization for storage of CO_2 in deep geologic formulations. U.S. Department of Energy, National Energy Technology Laboratory. Available via https://www.netl.doe.gov/File%20Library/Research/Carbon-Storage/ Project-Portfolio/BPM-SiteScreening.pdf. Accessed 28 Jan 2017

Saini D (2015) CO_2-Prophet model based evaluation of CO_2-EOR and storage potential in mature oil reservoirs. J Pet Sci Eng 134:79–86

Taber JJ, Martin FD, Seright RS (1997) EOR screening criteria revisited—part 1: introduction to screening criteria and enhanced recovery field projects. Soc Pet Eng. doi:10.2118/35385-PA

Yin M (2015) CO_2 miscible flooding application and screening criteria. MS Thesis. Missouri University of Science and Technology

Chapter 8
Reservoir Modeling of Simultaneous CO_2-EOR and Storage Projects

Abstract Reservoir modeling of simultaneous CO_2-EOR and storage projects is mainly performed for "history matching" the predicted behavior of injected CO_2 with measured behavior for ensuring it safe storage in subsurface. It can also be used to answer numerous management and optimization strategies related questions including effect of given variable on overall project performance, evaluation of a new CO_2 injection strategy, and means to enhance the storage capacity of CO_2 storage reservoirs.

8.1 Introduction

A combination of static geological modeling and dynamic reservoir simulation studies (i.e., reservoir modeling) is one of many tools that is used for designing, operation, and management of commercial simultaneous CO_2-EOR and storage projects. Reservoir modeling and simulation is used for predicting the movement and behavior of injected CO_2 and assist the operators in performing the risk analysis and monitoring needed to identify, estimate, and mitigate the risk arising from injection and permanent storage of injected CO_2. However, the access, affordability, and experience with commercial reservoir modeling software, often, pose significant challenges to the operators and stakeholders in performing modeling studies in timely and cost-effective manners.

Time to time, the public-private partnership has resulted in the development of reservoir modeling tools that are based on petroleum industry's experience and expertise in modeling and optimizing the performance of commercial CO_2-EOR projects. For example, in 1990s, Texaco Inc. with the support of the U.S. DOE, developed a CO_2 and waterflood performance software, namely the CO_2-Prophet Model (Dobitz and Prieditis 1994). Lately, a renewed push from the U.S. DOE's NETL, to model and study the commercial geologic CO_2 projects, has resulted in development of reservoir modeling tools like the COZView/COZSim software that allows the completion of integrated (technical and project economics) feasibility study in a relatively short period (NITEC website 2017). The recent additional

D. Saini, *Engineering Aspects of Geologic CO₂ Storage*, SpringerBriefs in Petroleum Geoscience & Engineering, DOI 10.1007/978-3-319-56074-8_8

refinements and parameter updates to the Kinder Morgan Inc.'s basic spreadsheet based analog economic scoping model for CO$_2$-EOR has extended the use of this model to many potential oil units while charactering incremental oil supply and CO$_2$ purchase demand for large regions (Cook 2012). The increased use of these tools (e.g., Saini 2015; NITEC LLC 2016; Agarwal et al. 2017; Advani and Ghaith 2014; Cook 2013) has resulted in rapid evaluation of favorable storage sites for consideration of deployment of commercial scale simultaneous CO$_2$-EOR and storage projects.

8.2 General Guidelines for Reservoir Modeling

Reservoir modeling serves several important roles in a geologic CO$_2$ storage project. It is used in evaluating the feasibility CO$_2$ storage in the subsurface, field tests' design, implementation, and analysis, and engineering and operating geologic CO$_2$ storage systems (Teletzke and Lu 2013). It also allows in answering some of the basic questions regarding the characterization of a potential CO$_2$ storage site (Lee 2016) including whether there is sufficient storage volume and whether it can be accessed (i.e. site capacity); whether suitable reservoir properties exist for sustained injection of CO$_2$ at economical industrial supply rates (i.e. site injectivity); whether the site is secure with negligible risk of unintended migration or leakage (i.e. site integrity); and can we ensure the injected CO$_2$ will remain in place (i.e. containment) (Andersen 2017).

The NETL's best practices for risk analysis and simulation for geologic storage of CO$_2$ (NETL 2011) discuss the ways in which behavior of injected CO$_2$ can be simulated and predicted by modeling the specific processes (thermal and hydrologic, chemical, mechanical, and biologic) in the subsurface via available codes. The reservoir modeling can enable the operator to "history match" the predicted location of the CO$_2$ and its measured location, thus allowing the operator to ensure that a project is safely storing CO$_2$ and can be safely closed once the site has reached a point of verified negligible risk (NETL 2011).

8.3 Additional Uses of Reservoir Modeling

Apart from history matching the predicted behavior of injected CO$_2$ with measured behavior for ensuring it safe storage in subsurface, reservoir modeling is a tool that can be used for answering numerous management and optimization strategies related questions. For example, answer of a question such as "by implementing a hybrid CO$_2$-EOR/storage scheme at the end of the CO$_2$-EOR operation, how much incremental CO$_2$ could be injected and how much incremental oil could be produced?" can only be answered by a reservoir modeling study (Jafari et al. 2013). Similarly, reservoir modeling can help in investigating the means that could result in enhances storage capacity of CO_2 storage reservoirs (e.g., Liu et al. 2013). A real field-scale evaluation and optimization of new CO_2 injection strategies such as the immiscible gas assisted gravity drainage (GAGD) process (Al-Mudhafar 2016) can only be realized via reservoir modeling.

However, the quality of such reservoir modeling results will rely on the quality of input data itself. Once, reservoir model is validated using field, laboratory, and/or pilot project(s) data, it can provide useful information on the influence of given variable(s) on overall project performance.

References

Advani D, Ghaith FA (2014) Applications of carbon capture and storage in enhanced oil recovery in UAE. Int J Eng Sci Inn Tech 3(4):502–511

Agarwal RK et al (2017) Carbon sequestration and optimization of enhanced oil and gas recovery. In: Agarwal AK et al (eds) Combustion for power generation and transportation, Springer Singapore, pp 401–432

Al-Mudhafar WJM (2016) Statistical reservoir characterization, simulation, and optimization of field scale-gas assisted gravity drainage (GAGD) process with uncertainty assessments. Dissertation, Louisiana State University, Baton Rouge

Andersen O (2017) Simplified models for numerical simulation of geological CO_2 storage. Dissertation, University of Bergen

Cook BR (2012) Wyoming's miscible enhanced oil recovery potential from main pay zones: an economic scoping study. Dissertation, University of Wyoming

Cook BR (2013) CO_2SCOPE^{TM} EORI's scoping model. Presented at the 19th annual CO_2 flooding conference, Midland, Texas, 11–13 Dec 2013. Available via www.co2conference.net/ wp.../10-Cook-UofW-CO2ScopeTM-CO2-Conf-2013.pdf. Accessed 12 Aug 2016

Dobitz JK, Prieditis J (1994) A stream tube model for the PC. Soc Pet Eng. doi:10.2118/27750-MS

Jafari A et al (2013) Transitioning of existing CO_2-EOR projects to pure CO_2 storage projects. Alberta innovates—technology futures. Available via http://www.ptac.org/attachments/1187/ download. Accessed 12 Feb 2017

Lee A (2016) CO_2 enhanced oil recovery (EOR) for capture and long-term underground storage of CO_2. Presented at COP22 side event, IETA Pavilion, Marrakesh, 15 Nov 2016. Available via http://cop22.co2geonet.com/media/6839/arthur_lee.pdf. Accessed 12 Feb 2017

Liu G et al (2013) Four-site case study of water extraction from CO_2 storage reservoirs. Energy Procedia 37:4518–4525

NETL (2011) Best practices for: risk analysis and simulation for geologic storage of CO_2. Best practice manual published by the National Energy Technology laboratory. Available via https:// www.netl.doe.gov/File%20Library/Research/Carbon%20Seq/Reference%20Shelf/BPM/BPM_ RiskAnalysisSimulation.pdf. Accessed 12 Feb 2017

NITEC LLC (2016) Frannie Tensleep in COZ. Available via http://www.uwyo.edu/eori/ technology-transfer/workshops/savage%20tensleep%20pres.pdf. Accessed 29 Jul 2016

NITEC LLC (2017) CO_2-EOR simulation software overview. http://www.nitecllc.com/COZSim_ COZView.html. Accessed 12 Feb 2017

Saini D (2015) CO_2-Prophet based evaluation of CO_2 EOR and storage potential in mature oil reservoirs. J Pet Sci Eng 134:79–86

Teletzke GF, Lu P (2013) Guidelines for reservoir modeling of geologic CO_2 storage. Energ Procedia 37:3936–3944

Chapter 9
Bringing Synergy Between EOR and Storage

Abstract An increased synergy between commercial CO_2-EOR and storage activities can further advance the geologic CO_2 storage as one of the core business activities of the petroleum industry. However, it is unlikely to be adopted by the petroleum industry without additional incentives. The petroleum industry has already capitalized on the ability to capture CO_2 at high-purity stationary sources such as natural gas processing, fertilizer production and an expansion of simultaneous CO_2-EOR and storage projects to the countries like China and India while creating additional pore space for storing large amounts of captured CO_2 will be the key for the future of geologic CO_2 storage for reducing GHG emissions around the world. The petroleum industry has skilled workforce that can support the initial growth of geologic CO_2 storage as a service industry in a future low-carbon economy, however, it is imperative that the technical managers, policy makers, and community leaders have an appreciation for various engineering aspects of geologic CO_2 storage projects so that they can devise necessary strategies needed to retrain existing workforce.

9.1 Status

In next 20 year, CO_2-EOR industry in the U.S. is on pace to store 1–2 billion tonnes of industrial CO_2 that would otherwise be vented (Wallace and Kuuskraa 2014), however, an increased synergy between EOR and storage activities can further advance the geologic CO_2 storage as one of the core business activities of the petroleum industry.

In 2015, the International Energy Agency (IEA 2015b) published a paper on combining EOR with CO_2 storage for profit. In this study, authors looked at the potential economic impact of modifying EOR in a manner that simultaneous CO_2-EOR and storage is economically interesting to petroleum industry. They also coined the term EOR+ for simultaneous CO_2-EOR and storage practices to emphasize on the storage objectives in addition to oil extraction. Based on the analysis for a hypothetical but representative field, they developed three models of

© The Author(s) 2017

D. Saini, *Engineering Aspects of Geologic CO₂ Storage*, SpringerBriefs in Petroleum Geoscience & Engineering, DOI 10.1007/978-3-319-56074-8_9

combining oil extraction with CO_2 storage for illustrating the technical and economic options. The three models were:

1. Conventional EOR+
2. Additional EOR+
3. Maximum Storage EOR+.

Obviously, Conventional EOR+ model is just continuing the current CO_2-EOR practices while fulfilling additional MVA/MMV needs. Additional EOR+ model relies on injection of larger amounts of CO_2 compared to current practices while improving the oil recovery factor. The Maximum Storage EOR+ model focuses on maximizing long-term storage of CO_2 in the oil reservoirs while achieving the same level of oil production as under the Advanced EOR+ model.

Other researchers such as Saini et al. (2013) have used detailed static geologic and dynamic simulation modeling of a commercial simultaneous acid gas EOR and storage project to explore the strategies like water extraction from an underlying water zone (aquifer) for additional gain in both oil recovery and storage capacity. The previously unknown depleted oil-bearing reservoirs, in the naturally water flooded formations, below the established oil-water contact in existing producing fields (residual oil zones (ROZs), appears to have potential for expending the scope of CO_2-EOR and storage while increasing the CO_2 demand well beyond the current volumes of CO_2 supply in the U.S. (Hill et al. 2013).

However, the practices like mentioned above and the models like Maximum Storage EOR+, even they are feasible, is unlikely to be adopted by the petroleum industry without additional incentives (IEA 2015b) such as commercialization of proof of concept technologies (e.g., extraction of formation water extraction for improving both oil recovery and storage capacity, CO_2 EOR and storage in ROZs), tax incentives for making the geologic CO_2 storage projects more attractive to potential operators even in low oil price regime, and sensible regulatory framework for promoting safe and effective execution of geologic CO_2 storage projects. Also, both industry and governmental entities have made some significant advancements and have taken new initiatives recently for promoting synergy between EOR and storage.

9.2 Future

The petroleum industry has already capitalized on the ability to capture CO_2 at high-purity stationary sources such as natural gas processing, fertilizer production. Recent launching of world's first commercial CO_2 capture facility at steam production plant to supply CO_2 for EOR project (ESI CCS project, Abu Dhabi, UAE) is the latest example of industry efforts to promote geologic CO_2 storage further. The Uthmaniyah CO_2-EOR demonstration project, launched by Saudi Aramco, national oil company of the Kingdom of Saudi Arabia shows the seriousness and proactive approach of industry leaders in bringing synergy between EOR and storage.

The People's Republic of China (PRC), largest energy consumer in the world of which 90% is fossil fuel-based, has started to conduct the first large scale demonstration project of CO_2-EOR and storage in Jilin oil field located in the Northeast China (Zhang et al. 2015). The PRC is also focusing on constructing new coal-fired power plants as CCS-ready (ADB 2015). Also, Australia and China have developed a plan for both CO_2-EOR and storage in the Liaohe oil field located in located in Liaoning Province, China (Kalinowski et al. 2013).

The Oil and Gas Climate Initiative (OGCI), whose member companies represent more than one-fifth of the world's oil and gas production have committed for $1 billion for reducing methane emissions in oil and gas production operations, accelerating the deployment of carbon capture, use, and storage (CCUS) via simultaneous CO_2-EOR and storage, improving industrial energy efficiency, and contributing to transportation efficiency (OGCI 2016).

The State of California, which recently entered in subnational global climate leadership memorandum of understanding with 33 countries (Under 2 Coalition 2016), has some history of CO_2 injection pilots in depleted oil field (Winslow 2012; Jeschke et al. 2000; MacAllister 1989) and its favorable climate policies may open the path for the deployment of simultaneous CO_2-EOR and storage projects in this U.S. State in near future.

Currently, there is no CO_2-EOR and storage in India, which is the fourth biggest contributor of GHG emissions in the world, however, recently reported R&D efforts in area of assessing the feasibility of CO_2-EOR and storage in mature oil fields (Ganguli et al. 2016) provides some hope that simultaneous CO_2-EOR and storage may become a reality soon in India as well.

9.3 Creation of Additional Storage Space

If all the associated GHG emissions for a CO_2-EOR project are taken in account, a net gain in CO_2 emissions to the atmosphere may result (e.g., Faltinson and Gunter 2013). Also, the depleted oil fields, are not capable of offsetting all of the GHG emissions that will result due to ultimate consumption of oil produced via CO_2-EOR operations. Hence, additional CO_2 storage space will be needed to offset any of the remaining GHG emissions as well as any of the produced CO_2 at the termination of simultaneous CO_2-EOR and storage projects. One attractive option is to inject CO_2 in the underlying water leg [aquifer or water zone below the established oil-water contact (OWC)], if present.

Second option will be the injection of CO_2 in candidate deep (>3000 ft) saline formations (dedicated CO_2 storage project) available nearby. An IEAGHG study (IEAGHG 2012) has investigated the feasibility of extracting formation water to increase amounts of CO_2 stored in both depleted oil fields as well as deep saline formations and the potential beneficial uses of produced formation water (e.g., agricultural irrigation). According to the study, ideal circumstances of relatively high quality reservoir water and highly stressed or limited regional water resources will need to coexist before beneficial use of extracted water may be considered. If proven commercial, CO_2 injection to extract formation water with or without CO_2-EOR component may prove to be boon to the regions around the world which are facing the serious draught (California) or are the oil rich regions with high water stress such as several middle-eastern countries and certain parts (Gujarat and Rajasthan) of India.

9.4 Workforce Training

Both the petroleum industry and the environmental consulting industry has a well-developed and highly competitive workforce that is needed for wide scale deployment of geologic CO_2 storage projects (National Research Council 2013). However, a lack of strong geosciences workforce resulting from recent decline in student recruitment into the geosciences could limit the availability of skilled workforce over the long term (National Research Council 2013).

According to an estimate by the National Research Council (2013), if large-scale implementation of CO_2-EOR is realized and geologic CO_2 storage projects quadruples from today's levels by 230, a workforce of 14,000–36,000 would be needed and in case of more accelerated growth, 35,000–90,000 workers might be needed for supporting geologic CO_2 storage projects. With some retraining, existing workforce can support the initial growth of geologic CO_2 storage as a service industry in future low-carbon economy. However, it is still necessary that the technical managers, policy makers, and community leaders have an appreciation and an improved understanding of various engineering aspects of geologic CO_2 storage projects for facilitating the retraining of existing workforce, creating opportunities for student recruitment into geosciences, and supporting graduate school research on geologic CO_2 storage topics so that any increased workforce demand for future geologic CO_2 storage project can be met in timely manner.

References

ADB (2015) Roadmap for carbon capture and storage demonstration and deployment in the People's Republic of China. Asian development bank. Available via https://www.adb.org/sites/default/files/publication/175347/roadmap-ccs-prc.pdf. Accessed 19 Dec 2016

Faltinson JE, Gunter B (2013) Net CO_2 stored in North American EOR projects. Soc Pet Eng. doi:10.2118/137730-PA

Ganguli SS et al (2016) Assessing the feasibility of CO_2-enhanced oil recovery and storage in mature oil field: a case study from Cambay basin. J Geol Soc India 88:273. doi:10.1007/s12594-016-0490-x

Hill B, Hovorka S, Melzer S (2013) Geologic carbon storage through enhanced oil recovery. Energy Procedia 37:6808–6830

IEAGHG (2012) Extraction of formation water from CO_2 storage. International Energy Agency (IEA) and Greenhouse Gas (R&D) Programme. Available via http://ieaghg.org/docs/General_Docs/Reports/2012-12.pdf. Accessed 27 Jan 2016

IEA (2015b) Storing CO_2 through enhanced oil recovery (combining EOR with CO_2 storage (EOR+) for profit. International energy agency. Available via https://www.iea.org/publications/insights/insightpublications/Storing_CO2_through_Enhanced_Oil_Recovery.pdf. Accessed 22 Jan 2017

Jeschke PA, Schoeling L, Hemmings J (2000) CO_2 flood potential of California oil reservoirs and possible CO_2 sources. Soc Pet Eng. doi:10.2118/63305-MS

Kalinowski A et al (2013) China-Australia capacity building program on the geological storage of carbon dioxide—results from Phase I. Energy Procedia 37:7299–7309

MacAllister DJ (1989) Evaluation of CO_2 flood performance: North Coles Levee CO_2 pilot, Kern County. California Soc Pet Eng. doi:10.2118/15499-PA

National Research Council (2013) Emerging workforce trends in the U.S. energy and mining industries: a call to action. The national Academies Press, Washington DC

OGCI (2016) Taking action: accelerating a low emissions future. Oil and gas climate initiative. Available via http://www.oilandgasclimateinitiative.com/news/2016-report-taking-action-accelerating-a-low-emissions-future. Accessed 23 Dec 2016

Saini D et al (2013) A simulation study of simultaneous acid gas EOR and CO_2 storage at Apache's Zama F Pool. Energy Procedia 37:3891–3900

Under 2 Coalition (2016) The memorandum of understanding (MOU) on subnational global climate leadership. Available via http://under2mou.org/the-mou/. Accessed 28 Jan 2017

Wallace M, Kuuskraa V (2014) Near-term projections of CO_2 utilization for enhanced oil recovery. U.S. Department of Energy, National Energy Technology Laboratory. Report no. DOE/NETL-2014/1648

Winslow D (2012) Industry experience with CO_2 for enhanced oil recovery. Workshop on California opportunities for CCUS/EOR. Available via www.usea.org/sites/default/files/eventpresentations/Califonia%20CCUS_EOR%20Workshop%20Jun%2027%202012%20Agenda.pdf. Accessed 26 Jan 2013

Zhang L et al (2015) CO_2 EOR and storage in Jilin oilfield China: monitoring program and preliminary results. J Pet Sci Eng 125:1–12

Printed in the United States
By Bookmasters